How
WE
GOT
TO
NOW

Students experimenting with batteries and electricity at a Washington, D.C., high school, circa 1899.

STEVEN JOHNSON

HOW WE GOT TO NOW

Six Innovations That Made ┈┈┈┈┈►
◄┈┈┈┈┈ THE MODERN WORLD

adapted by Sheila Keenan

VIKING

VIKING

An imprint of Penguin Random House LLC

375 Hudson Street

New York, New York 10014

First published in the United States of America by Viking,
an imprint of Penguin Random House LLC, 2018

LIBRARY OF CONGRESS CATALOGING-IN-PUBLICATION DATA
Names: Johnson, Steven, 1968– author.
Title: How we got to now : six innovations that made the modern world /
by Steven Johnson.
Description: [Young Readers edition]. | New York : VIKING, Published
by Penguin Group, 2018. | Audience: Ages 10+ | Audience: Grades 5 up. |
Identifiers: LCCN 2017060707 (print) | LCCN 2017061595 (ebook) | ISBN
9780425287804 (ebook) | ISBN 9780425287781 (hardcover) | ISBN
9780425287798 (trade pbk.)
Subjects: LCSH: Technology—Social aspects—Juvenile literature. |
Inventions—Social aspects—Juvenile literature. | Technology and
civilization—Juvenile literature.
Classification: LCC T14.5 (ebook) | LCC T14.5 .J642 2018 (print) |
DDC   303.48/3—dc23
LC record available at https://lccn.loc.gov/2017060707

Manufactured in China   Set in Excelsior LT Std
Book design by Jim Hoover

10   9   8   7   6   5   4   3   2   1

For Dean

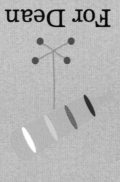

CONTENTS

*Construction workers build a skyscraper high above
New York City's Fifth Avenue, circa 1908.*

Introduction

················▶

How was your day?

Did you wake up in a room comfortably cool or warm? Did you shower in a strong, steady stream of water? Could you see to get dressed when it was still dark out? Did you gulp down a carton of milk at lunch even though there isn't a cow in sight? Could you tune out a pesky sibling because your earbuds were firmly implanted?

So the day was OK. Nothing remarkable—until you stop to think about the fact that you can control temperature, light, sound, water; basically everything in your environment.

Today's technology and innovation move at dizzying speeds.

Most of us in the developed world don't give a passing thought to how amazing it is that we enjoy luxuries like drinking water that won't kill us because of bacterial disease, artificial light that stretches the day into night, or air-conditioning that lets many of us live comfortably in climates that would have been intolerable just sixty years ago. We don't really even consider these luxuries at all. But who made all this possible?

No one.

As in, *no one* person.

If you want to understand how big ideas truly changed the world, you need to get rid of the myth of the "Eureka!" moment. Brilliant inventions don't come about because some solitary genius is smarter than everyone else. Ideas are fundamentally *networks* of other ideas. We take the tools and metaphors and concepts and scientific understanding of our time, and we remix them into something new. But if you don't have the right building blocks, you can't make the breakthrough, however brilliant you might be.

Our lives are surrounded and supported by a whole class of objects and systems that are powered by the creativity of thousands of people who came before us: inventors and hobbyists and reformers who steadily hacked away at a problem or refused to let go of what I like to call a "slow hunch," an idea that comes into focus over decades, not seconds.

Ordinary people doing extraordinary things.

These tinkerers and dreamers are the main characters in our story. Their ingenious plans rarely transformed the world immediately. Mostly they had hunches, thoughts that were vague, even wrongheaded, but hinted at something bigger. And eventually, collectively, these ideas ushered in revolutions in the way we live.

Inventions and scientific discoveries tend to come in clusters at specific moments in history, when a handful of geographically dispersed investigators stumble independently onto the very same thing. The electric battery,

An electric car's batteries getting charged in Detroit, Michigan, circa 1919. Electric cars were originally developed in the 1890s.

the telegraph, the steam engine, and the digital music library were all independently invented by multiple people in the space of a few years. Scientists and scholars have tracked hundreds of these simultaneous inventions.

The innovations in this book belong to everyday life, not science fiction: a glass lens, air-conditioning, a sound recording, a cup of clean tap water, a wristwatch, a lightbulb. This is a history worth telling, in part because it allows us to see a world we generally take for granted with fresh eyes. But the other reason to write this kind of history is that these innovations have set in motion a much wider array of changes in society than you might expect.

Innovations usually begin with an attempt to solve a specific problem, but once they get into circulation, they end up triggering other changes that would have been extremely difficult to predict. We like to think we are decision-makers in charge of our world. And often change does come about through the conscious planning and actions of political or military leaders, or artists, or scientists or inventors, or voters or protest movements, any of which may deliberately bring about some kind of new reality. But social transformations are not always the direct result of deliberate human choices. In some cases, ideas and innovations take on a life of their own—and they spark changes that were not part of their creators' vision. You might not think that the invention of air-conditioning would change American politics, or that Gutenberg's printing press would lead directly to the creation of telescopes and microscopes, but that's the unlikely way that important innovations can sometimes shape the world.

Innovations can have mixed consequences as they become more widespread. Cars move us more efficiently through space than horses did, but are they worth the impact on the environment? Cell phones and texting mean almost instant access to people and information, but how does that affect live conversations and other social skills, sharing public spaces, and even driving safety?

I should mention two additional elements of the book's focus: the "we" in this book, and in its title, is largely the "we" of North Americans and Europeans. (The story of how China, India, or countries in the Middle East or South America got to now would be a different one, and every bit as fascinating.) Certain critical experiences—the rise of the scientific method, industrialization—happened in Europe *first*, and have now spread across the world. (*Why* they happened in Europe first is of course one of the most interesting questions of all, but it's not one this book tries to answer.)

Also, while a few brilliant and innovative women appear in this book—like the first computer programmer, Ada Lovelace, and the entrepreneur Annie Murray—most of the stories date back to times when women were actively discouraged from pursuing careers as scientists, inventors, or entrepreneurs. Because of that, the vast majority of the innovators profiled in these pages are men. Thankfully, advances in gender equality over the last few decades have made it easier for women to make scientific discoveries and create world-changing devices, though of course there is still work to do to reach a completely level playing field. I have no doubt that future historians writing a new version of this book fifty years from now could feature only female innovators if they wanted.

The story I'm telling is "long-zoom" history: traditionally, we look at history through the narratives of individuals or nations, but on some fundamental level, these boundaries are too limiting. History happens on the microscopic level of atoms, on the vast level of planetary climate change, and on all the levels in between. To get the story right, we need a framework that does justice to all those different levels. For example, to understand why transparent glass came to change so much of the modern world, we have to peer into the subatomic properties of silicon dioxide, the material that glass is made from—and we have to zoom out to see the impact of the glassmaking industry on the city of Venice.

From the places we live, to the food we eat, to what we produce and consume, to how we inform or entertain ourselves, I want to show you how these seemingly unconnected worlds are linked by the unsung heroes whose questions, curiosity, and doggedness led to the inventions and chain reactions that shape our modern world.

In other words: how we got to now.

Marin County, California
August 2017

Glass

••••••••••••••••••••••••••••••►

To say that something hot happened in a desert doesn't sound all that remarkable. But what we're talking about was *very* hot—and the impact was extraordinary.

Roughly twenty-six million years ago, grains of silica in the sands of the Libyan Desert melted and fused under an intense heat that must have been at least 1,000 degrees Fahrenheit (538 degrees Celsius). Many scientists think a comet collision and explosion caused the event. When those superheated grains of sand cooled down below their melting point, a vast stretch of the Libyan Desert was coated with a layer of . . . *glass.*

About ten thousand years ago, give or take a few millennia, someone traveling through the desert stumbled across a large

The skyline of Dubai, United Arab Emirates, features incredible skyscrapers, especially Burj Khalifa, the world's tallest building, which includes 1.8 million feet of glass.

fragment of this glass. It circulated through the markets and social networks of early civilization, and ended up as the centerpiece of a brooch, carved into the shape of a scarab beetle. The glass sat there undisturbed for four thousand years, until archeologists unearthed it in 1922 in the burial chamber of the pharaoh Tutankhamun, more commonly known as King Tut.

Silica, or silicon dioxide, is a chemical compound that is abundant on our planet. More than 50 percent of Earth's crust is made up of this element, which plays almost no role in the natural metabolisms of carbon-based life-forms (like us) or in technologies that rely on fossil fuels and plastics. One reason evolution didn't find much use for silicon dioxide is that most of the really interesting things about this substance don't appear until it's in temperatures of over 1,000 degrees Fahrenheit. That kind of heat doesn't happen on the surface of the planet without the right technology—which we got once people invented the furnace. By learning how to generate extreme heat in a controlled environment, we unlocked the molecular potential of silicon dioxide, and figured out how to create glass. Soon, the way we saw the world and ourselves was transformed.

Made of Glass

Glass made the transition from ornament to advanced technology during the height of the Roman Empire in the first and second centuries. That's when glassmakers figured out ways to produce sturdier and less cloudy material than the natural glass used in King Tut's scarab. Roman craftsmen shaped the melted silica into drinking and storage vessels and windowpanes, the very first ones built.

Jump ahead a millennium and east to Constantinople, a wealthy city that is now Istanbul, Turkey. A holy war that caused the siege and destruction of this city in 1204 was one of those historical quakes that send tremors of

This pectoral, a type of jewelry, was unearthed in the tomb of Pharaoh Tutankhamun.

influence rippling across the globe. It also sent a small community of glass-makers from Turkey westward across the Mediterranean; they landed in the Republic of Venice, a prosperous city on the shores of the Adriatic Sea.

In the thirteenth century, Venice was the most important commercial trading hub in the world. The Turkish glassmakers who settled there and stoked their blazingly hot furnaces created new luxury goods for merchants to sell around the globe. They also accidentally burned down neighborhoods.

In 1291, in an effort to retain the profitable skills of the glassmakers and protect public safety, the Venetian government ordered the Turks to relocate a mile across the Venetian Lagoon to the island of Murano. Unwittingly, they created what we now call an "innovation hub": a place like Silicon Valley, where new ideas and technologies tend to prosper.

Economists have a term for this phenomenon: "information spillover." Pack people together and ideas have a natural tendency to flow from mind to mind. Murano was densely populated, which meant that new ideas about glassmaking spread quickly, especially since many of the glassmakers were related. The success of Murano was shaped by sharing as much as by competition.

One member of that creative community, Murano glassmaker Angelo Barovier, clearly made one of the major breakthroughs in glass. And "clear" was what he sought. After years of trial and error with different substances, Barovier burned saltwort, a mineral-rich plant he had imported from Syria hundreds of miles away; extracted the minerals from the ashes; and added them to molten glass. When the mixture cooled, it created an extraordinary type of glass: you could see through it. This was the birth of modern glass.

Roman glass from the second century. Many ancient civilizations had the ability to make glass items, but they were opaque and often colored.

By the 1300s, Murano had become known as the Isle of Glass, and its ornate vases and other exquisite glassware became status symbols throughout Western Europe. Glassmakers still work there today, and many are direct descendants of the original families that emigrated from Turkey.

Today, of course, we take it for granted that glass is a transparent material, and we're so accustomed to seeing glass everywhere in our world that we don't even think of it as a technological advance. But eight hundred years ago, the ability to create transparent glass made Murano one of the most technologically advanced places on the planet, and Barovier's breakthrough would turn out to be far more important than he could have realized.

As we now know, most materials absorb the energy of light. But because of its composition, silicon dioxide allows light to pass through. That's why glass is transparent. It can also be used to bend, distort, or even break up light waves. This property of glass turned out to be even more revolutionary than simple transparency.

Seeing More Clearly

During the twelfth and thirteenth centuries, monks in European monasteries labored over religious manuscripts. The pages were intricately lettered and drawn, and the monks often worked in flickering candlelight. Many of them started using curved chunks of glass as reading aids. The glass worked as a bulky magnifier over the page.

No one is sure exactly when or where it happened, but in Northern Italy, medieval glassmakers took that innovation another step forward. They shaped glass into two small disks that bulged in the center, placed each disk in a frame, and joined the frames together at the top. Behold, the world's first spectacles!

For several generations, this ingenious device was used almost exclusively by monks and scholars. At the time, the vast majority of people were illiterate; they had pretty much no need to decipher tiny shapes like letters as part of their daily routine. Spectacles remained rare and expensive.

From 1342, this is the earliest known image of a monk with glasses. These early spectacles were called roidi da ogli, *"disks for the eyes." Thanks to their resemblance to lentils—*lentes *in Latin—the disks came to be called "lenses."*

A German metalworker changed all that.

Johannes Gutenberg invented the printing press in the 1440s, modeling it after a wooden screw press used to crush grapes for wine. He produced reusable individual letters made of metal (movable type) and started printing. Gutenberg's famous printing press is a great example of adapting an existing technology for a whole new outcome in a totally different field.

Thanks to him, for the first time printed books were relatively cheap and portable. This triggered a surge of literacy that opened people's eyes—in more ways than one. A massive number of people became aware that they were farsighted; they couldn't see things up close—like a printed page—very clearly. They needed glasses.

The Gutenberg Bible, published in the 1450s, was the first major book printed with movable type. This volume belongs to the New York Public Library's collection.

CONSPICILLA.

Inuenta conspicilla sunt, quæ luminum   Obscuriores detegunt caligines.

This print (circa 1600) shows a market with a spectacles seller (left) *and a bookseller* (right) *and several people wearing glasses.*

Within a hundred years of Gutenberg's invention, thousands of spectacle makers around Europe were thriving. Glasses were the first piece of advanced technology that ordinary people regularly wore on their bodies since the invention of clothing way back in Neolithic times.

Europe was not just awash in lenses, but also in *ideas* about lenses. For the first time, the properties of silicon dioxide were about to be harnessed not just to improve what we could see with our own eyes, but to see things that transcended the natural limits of human vision.

Better Living
Through Lenses

In the 1590s in the small town of Mid-delburg in the Netherlands, father-and-son spectacle makers Hans and Zacharias Janssen experimented with placing two lenses in line with each other, which magnified the objects they observed.

The Janssens invented the microscope.

Within seventy years, the British scientist Robert Hooke made a most influential discovery using this invention. He looked at a thin sheaf of cork through a microscope lens and observed that it was made of "pores, or cells, [that] were not very deep, but consisted of a great many little Boxes." Hooke had just identified one of life's fundamental building blocks: the cell. A revolution in science and medicine was set in motion.

The microscope took nearly three generations to produce truly trans-formative science, but another instru-ment with glass lenses spurred change more quickly. Twenty years after the invention of the microscope, a cluster

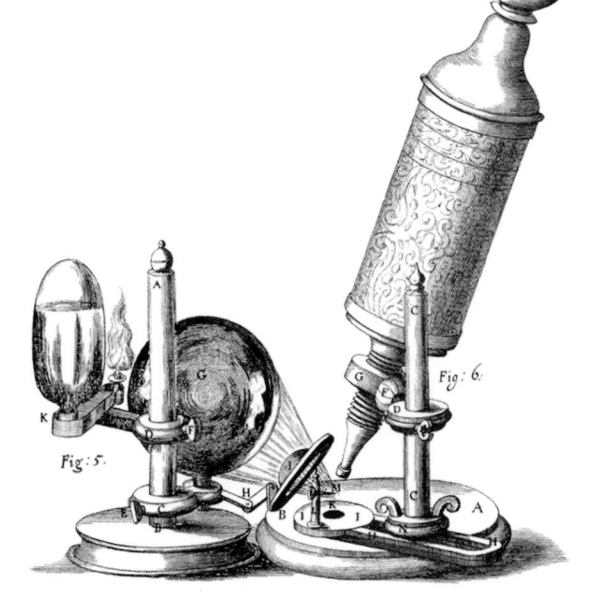

Robert Hooke's microscope, 1665. Over time, microscopes revealed invisible colonies of bacteria and viruses, which in turn led to modern vaccines and antibiotics.

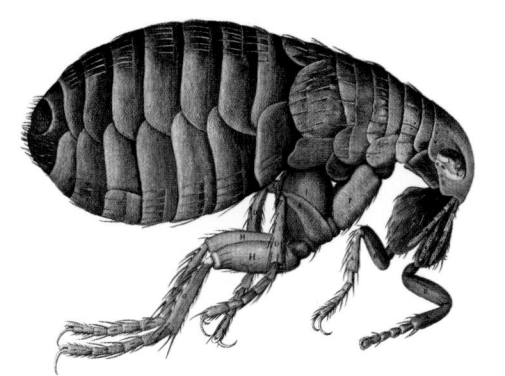

Hooke analyzed fleas, flies, wood, leaves, and even his own frozen urine through his microscope, and then drew what he saw in the groundbreaking book Micrographia *published in 1665.*

of Dutch lens makers more or less simultaneously invented the telescope. When the Italian scientist Galileo Galilei learned of this miraculous new device "for seeing things far away as if they were nearby," he modified its design to reach a magnification of ten times normal vision. In January of 1610, Galileo used his telescope to observe that moons were orbiting Jupiter. This was the first real challenge to the deeply held belief that all heavenly bodies circled the Earth.

Several hundred years later, in the nineteenth and twentieth centuries, the lens continued to have a huge impact on society. The camera lens helped photographers focus beams of light on specially treated paper to capture images. Film cameras and projectors also used lenses to record and then show "motion pictures" for the first time. People loved going out to the movies and, later, staying in to watch television, which got its start in the 1940s, when inventors discovered that coating glass with phosphor and firing electrons at it created images.

Think about all the essential components of modern life that have their roots in transparent glass: the medical

A portrait of Galileo Galilei. Gutenberg's printing press helped scientists like Galileo circulate their ideas. But Galileo's work contradicted the Roman Catholic Church's teachings, and he ended up imprisoned in Florence, Italy.

breakthroughs that come from studying cells and microbes through micro-scopes; photographs, television shows, and blockbuster movies; the wind-shields of automobiles and airplanes; glass skyscrapers. How different would the world be today if the Venetians had failed to invent transparent glass, or if silicon dioxide didn't allow light to pass through it in the first place?

All these inventions depend on the unique ability of glass to transmit and manipulate light. But glass has another physical property that the master glassmakers of Murano and the Renaissance lens makers had failed to exploit. This aspect of glass would also transform modern life—once a man with a crossbow discovered it.

Ready, Aim, Fire!

Charles Vernon Boys was a physicist at London's Royal College of Science in the 1880s. He had a gift for designing and building scientific instruments. In 1887, Boys wanted to create a very fine shard of glass to measure the effects of extremely small forces on objects. His idea centered on using a thin fiber of glass to help create his measurement tool. But where to get that fiber?

A new form of measurement almost always involves making a new measuring tool. Boys took an unusual approach to develop his tool: he brought a crossbow into his lab. He created lightweight arrows (or bolts) for the weapon. Then, using sealing wax, he attached the end of a glass rod to a bolt, heated the glass rod until it softened, and fired!

The bolt hurtled toward its target, pulling a thin strand of fiber from the molten glass still clinging to the crossbow. In one of his shots, Boys produced a thread of glass that stretched almost ninety feet long. Even more astonishing, the glass fiber was as durable as a strand of steel the same size.

For thousands of years, people had made things with glass because it was beautiful and transparent, but they had to work around the fact that it was

Charles Vernon Boys in a laboratory, 1917.

fragile. Boys's sure shot with his crossbow suggested that there was a whole other way to think about this amazingly versatile material: using glass for its *strength*.

In the 1930s, glass fibers began to be mass-produced. When plastic resin was added, a whole new construction material became available: fiberglass. Fiberglass is strong, bendable, and now ubiquitous. It's used in building insulation, clothes, surfboards, yachts, helmets, computer circuit boards—pretty much everywhere. The blades of wind turbines, which are changing the possibilities of alternative energy, are made of fiberglass. So is the fuselage of Airbus's A380, the largest commercial aircraft in the skies. A composite of aluminum and fiberglass makes the plane much more resistant to fatigue and damage than traditional aluminum shells.

During the first decades of innovation with glass fibers, this emphasis on strength versus transparency made sense. It was useful to allow light to pass through a windowpane or a lens, but why would you think about how it passes through a fiber not much bigger than a human hair? The transparency of glass fibers became an asset once we began thinking of light as a way to encode digital information.

Turn on the Laser Beam

In 1970, researchers at Corning Glass Works, the Murano of modern times, developed a type of glass so extraordinarily clear that if you produced a block of it the size of a bus, the glass would still be just as transparent as a normal windowpane. (We can now make a glass block a half mile long with the same clarity.) Scientists at Bell Labs then took fibers of this super-clear glass and shot laser beams down the length of them, in fluctuating optical signals that corresponded to binary code. (Binary code uses zeros and ones to represent letters, numbers, or other characters in an electronic device.

It's the language used in computing and telecommunications.) This combination of two seemingly unrelated inventions—the concentrated, orderly light of lasers, and hyper-clear glass fibers—came to be known as fiber optics. Using fiber-optic cables was much more efficient than sending electrical signals over copper cables, particularly for long distances. Light allows much more bandwidth and is far less susceptible to noise and interference than electrical energy. Today, the backbone of the global Internet is built out of fiber-optic cables. That's right: the World Wide Web is woven together out of threads of glass.

Think of that iconic, twenty-first-century act: snapping a selfie on your phone, and then uploading the image to an app, where it circulates to other people's phones and computers all around the world. Then think of how many ways glass supports this entire event: we take pictures through glass lenses, store and manipulate them on circuit boards made of fiberglass, transmit them around the world via glass cables, and enjoy them on screens made of glass. It's silicon dioxide all the way along the chain. And in fact, the connection between self-portraits and glass has a long history.

I See, Therefore I Am

Before 1400, European artists painted landscapes, royalty, religious scenes, and many other subjects. But they didn't paint themselves. Self-portraiture—the original selfie—was the direct result of yet another technological breakthrough in our ability to manipulate glass.

In a new innovation, glassmakers in Murano coated the back of clear glass with a metal mixture of tin and mercury to create a shiny and highly reflective surface: a mirror. This was literally a revelation. Before mirrors came along, most people had only glimpsed themselves in distorted reflections in water or polished metals; they went through life without ever seeing truly accurate self-images. Imagine never being entirely sure what you really looked like. That was the reality until the invention of accurate mirrors.

Just as the glass lens was extending our vision to the stars and to microscopic cells, glass mirrors showed us ourselves for the first time. The impact on society was profound. The mirror played a direct role in allowing artists to paint self-portraits and invent perspective as a formal device in visual art. Shortly thereafter, a fundamental shift toward individuality occurred in European culture. Once you saw yourself, you were more likely to think of yourself as the center in relation to the state, law, economics, even your god. Because of this new way of seeing, laws were increasingly oriented around individuals, bringing about a new emphasis on human rights and individual liberties. Many forces converged to make this shift possible, and mirrored glass was one of them.

Glass helped invent our modern sense of self; now it's helping us explore worlds beyond ourselves—from atop a volcano.

Looking Out at the Universe

Mauna Kea, a dormant volcano on Hawaii's Big Island, rises almost fourteen thousand feet above sea level and extends almost another twenty thousand feet down to the ocean floor. The landscape at its summit is rocky and barren. Even clouds generally sit several thousand feet below the volcano's peak, where the air is dry and thin. On the summit, you are as far from the continents of Earth as you can be while standing on land, which means the atmosphere around Hawaii is undisturbed by the turbulence of the sun's energy bouncing off or being absorbed by large, varied landmasses. It's pretty much the most stable atmosphere on the planet. All of which makes this volcano's peak a perfect location for stargazing.

A laser beam shoots out from the Keck II telescope on a clear, moonlit night.

Mauna Kea is crowned by thirteen observatories, massive white domes scattered across the red rocks like some gleaming outpost on a distant planet. One of the domes is the W. M. Keck Observatory, which houses the largest, most powerful optical telescopes on Earth. The Keck twin telescopes do not rely on lenses to do their magic. To capture as much light as possible from distant corners of the universe and learn something about stars and galaxies, you would need lenses the size of a pickup truck; at that size, glass becomes difficult to physically support and introduces inevitable distortions into the image. So the scientists and engineers of Keck employed another tool to capture extremely faint traces of light: the mirror.

The Keck Observatory on Mauna Kea.

Each of the Keck telescopes has thirty-six hexagonal mirrors that work together as a single reflective surface thirty-three feet in diameter. Incoming starlight bounces up to a second mirror, then focuses down and is captured by a set of instruments that processes the images and visualizes them on a computer screen. Even in Mauna Kea's thin, ultra-stable atmosphere, small disturbances can blur the captured images. The observatory uses an ingenious system called "adaptive optics" to correct the vision of the telescopes. Lasers are beamed into the night sky above Keck, creating an artificial star. That false star acts as a reference point because scientists know exactly what the laser would look like if there were no atmospheric distortion. So they are able to measure any existing distortion by comparing the "ideal" laser image and what the Keck telescopes register. Based on this information, computers generate instructions so the mirrors of the telescope flex slightly and adjust for the exact distortions in the skies above Mauna Kea on a given night. It's as if you were farsighted and suddenly put on glasses to read.

When we look through the telescopic mirrors of Keck, we are looking into the distant past, because the objects we see are galaxies and supernovas that can be billions of light-years away. Once again, glass has extended our vision, not just down to the invisible world of cells and microbes, or through the global connectivity of a smartphone, but all the way back to the early days of the universe. Glass started out as trinkets and empty vessels. A few thousand years later, perched above the clouds at the top of Mauna Kea, it has become a time machine.

Surrounded by Silicon

Maybe you're wearing glasses to see this book clearly. Or you're looking at this page on a smartphone or tablet. Perhaps you're multitasking, reading my words *and* watching a video on YouTube. Wherever you are and whatever you're doing, there are probably a hundred objects within reach that depend on silicon dioxide for their existence, and even more that rely on the element silicon itself: the panes of glass in your windows or skylights, the lens in your phone's camera, the screen of your computer, everything with a microchip or digital clock. Glass changed the way we see and experience the world and broadened our understanding of humanity. We're now able to see in focus, to see ourselves, to see the invisible, to see beyond our world, and to share our vision with others. No material on Earth mattered more to those conceptual breakthroughs than glass.

Cold

·······························▶

Imagine you were one of the many human beings living in tropical climates before the 1800s; as strange as it sounds to us now, you would never have experienced anything frozen in your entire life. Not just snow—you would never have seen even an ice cube. And worse, no ice cream!

People have been doing imaginative things with ice in frozen parts of the world forever. Today, for example, you can check into a hotel where your room is ten below freezing and sleep on a bed of ice. But it was only a few hundred years ago that we realized how we could use ice and cold in *warmer* climates. Now you can live or even go snow skiing in the desert. This cold revolution began with a simple desire for an ice-cold drink on a hot summer's day.

An ice hotel in Jukkasjärvi, Sweden, 2014. The hotel is rebuilt every winter, and this version included one hundred rooms and a secret garden. Temperatures inside hover at 23 degrees Fahrenheit (−5 C).

Tropical Ice

In the early summer months of 1834, a three-masted ship named the *Madagascar* sailed into the port of Rio de Janeiro, Brazil. Its hull was filled with a most unlikely cargo: a frozen New England lake.

The *Madagascar* and her crew were in the service of an enterprising and dogged Boston businessman named Frederic Tudor. For most of his adult life, Tudor was a dreadful failure. But he refused to let go of a slow hunch he had about ice, even when it cost him his sanity, his fortune, and his freedom. His tenacity paid off: Frederic Tudor is now known as "the Ice King."

As a well-to-do young Bostonian, Tudor had long enjoyed the frozen water from the pond on Rockwood, his family's country estate. Aside from its natural wintery beauty, the pond ice was extremely useful when it came to keeping things cold all year long. Like many wealthy families in northern climes, the Tudors stored blocks of frozen water. Inside their icehouse, huge, two-hundred-pound ice cubes sat solidly cold and unmelted. When the hot summer months arrived, icehouse owners could have slices chipped off these blocks to freshen drinks, make ice cream, or cool bathwater during a heat wave.

Frederic Tudor knew from personal experience that a large block of ice could last well into the depths of summer if it was kept out of the sun. That knowledge planted the seed of an idea in his mind. A trip to the tropics cultivated that seed.

In 1801, seventeen-year-old Frederic and his older brother John, who suffered from a painful knee ailment, set sail on a voyage to the Caribbean in hopes that the climate would help John's health. The plan did not work out as expected: in Havana, Cuba, the Tudor brothers were quickly overwhelmed by the hot and humid weather. They soon sailed back to the United States, but the muggy summer heat came with them. Just six months later John died.

In the oppressive, inescapable humidity of the tropics, Tudor thought about how delicious it would be to sip a cold drink. Surely if he could

Frederic Tudor was brisk, confident, and almost comically ambitious. Three decades after his first failed voyage, Tudor wrote in his journal: "The business is established. It cannot be given up now and does not depend upon a single life. Mankind will have the blessing for ever whether I die soon or live long."

Ice harvesting in Massachusetts, circa 1850. Boys cleared the snow with shovels and then men hand cut rough, large blocks with saws. The invention of a dual-blade, horse-drawn ice cutter, shown above, made the process easier.

somehow transport ice from the frozen north to the West Indies, there would be a huge market for it. This was a radical idea, but the history of global trade had demonstrated that vast fortunes could be made by transporting a commodity that was plentiful in one place to another where it was scarce. Frederic thought ice fit this equation perfectly: nearly worthless in Boston, it would be priceless in the Caribbean.

Several aimless years passed. Frederic Tudor's ice trade idea remained nothing more than a hunch. Still, he held on to his seemingly preposterous notion and eventually shared it with another brother, William. Frederic began taking notes in a journal he called his "Ice House Diary." The diary shows supreme confidence in his scheme: "In a country where at some seasons of the year the heat is almost unsupportable," he wrote, "ice must be considered as out doing most other luxuries." He was firmly convinced that the ice trade would bring the Tudor brothers "fortunes larger than we shall know what to do with." (William was apparently less convinced of the promise of his brother's scheme.) Frederic seems to have given less thought to how to transport the ice to the Caribbean—or what to do with it when it arrived.

However deluded his idea might have seemed, Tudor did have the means to put his plan in motion. He had enough money to hire a ship, and an endless supply of free ice, manufactured by Mother Nature. And so, in November 1805, he dispatched his brother and cousin to the Caribbean island of Martinique in the West Indies. Their job was to secure proper storage and a local buyer who wanted exclusive rights to sell the ice that Frederic would bring several months later.

While waiting for word from his envoys, Tudor bought a sailing ship called the *Favorite* for $4,750 and began harvesting ice in preparation for the roughly two-thousand-mile journey. In February 1806, Tudor and a full cargo of Rockwood ice set sail from Boston Harbor, to the amusement of the press. As the *Boston Gazette* printed, "We hope this will not prove a slippery speculation."

The three-week voyage was beset by weather-related delays, but the ice survived the journey to Martinique in remarkably good shape. The problem turned out to be one that had never occurred to Tudor. The islanders had no interest in his exotic frozen bounty. They simply had no idea what to do with it.

Sometimes an object's uniqueness makes its use hard to figure out. Tudor had assumed the sheer novelty of his ice blocks would be a point in his favor. Instead, the frozen water just received blank stares. Overall indifference to the idea of trading in ice had prevented William Tudor from lining up an exclusive buyer for the cargo. Even worse, he had failed to find a place to store it.

Tudor's investment melted in the tropical heat at an alarming rate. He posted flyers around town with instructions for how to carry and preserve the ice. He impressed a few locals by making ice cream, an unusual treat so close to the equator. But sales were poor. In his diary, Tudor noted that he had lost nearly four thousand dollars with his tropical misadventure. Still, he kept sending ice ships to the Caribbean, though few on the islands were eagerly awaiting his cargo. Meanwhile, the Tudor family fortunes collapsed, and shipwrecks and embargoes made things worse. By 1813, Frederic was broke and in debtor's prison.

But Tudor persevered. His New England home gave him one crucial advantage, beyond the ice itself. Unlike the American South, with its sugar plantations and cotton fields, the northeastern states were largely devoid of natural resources that could be sold elsewhere. This meant that ships tended to leave Boston Harbor empty, heading to the West Indies to fill their hulls with valuable cargo before returning to the wealthy markets of America's eastern seaboard. Paying a crew to sail an empty ship was effectively burning money. Any cargo was better than nothing, which meant that Tudor could negotiate cheaper rates for himself by loading his ice onto what would have otherwise been an empty ship. He could then avoid the costs of buying and maintaining his own vessels.

Part of the beauty of ice, of course, was that it was basically free: Tudor needed only to pay workers to carve blocks of it out of a frozen body of water. New England's economy generated another product that was considered equally worthless: sawdust, the primary waste product of lumber mills. After years of experimenting with different options, Tudor had discovered that sawdust made a brilliant insulator for his ice. Blocks layered on top of each other with sawdust separating them would last almost twice as long as unprotected ice. This was Tudor's frugal genius: he took three things that the market had effectively priced at zero—ice, sawdust, and an empty ship—and eventually turned them into a flourishing business.

He also learned from the first, catastrophic trip to Martinique and tinkered with multiple icehouse designs to insulate his cargo from the tropical heat. He finally settled on a double-shelled structure that used the air between two stone walls to keep its interior cool.

Tudor didn't understand the molecular chemistry of his shipping and storing methods, but both the sawdust and the double-shelled architecture revolved around the same principle. Ice melts when it pulls heat from the surrounding environment. That process happens on the outside surface of the ice, and if you put a buffer between the ice and the heat, you can slow down the melting. Air is a good buffer because it doesn't conduct heat efficiently. There was plenty of air between the walls of Tudor's double-shelled icehouses. And his sawdust packaging on the ships ensured that there were pockets of air between the wood shavings to keep the ice insulated.

By 1815, Tudor had finally assembled the key pieces of the ice puzzle: harvesting, insulation, transportation, and storage. Still pursued by his creditors, he began making regular shipments to a state-of-the-art icehouse he had built in Havana, where an appetite for ice cream had been slowly growing. Fifteen years after his original hunch, Tudor's ice trade finally turned a profit. By the 1820s, he was shipping and storing frozen New England water all over the American South. By the 1830s, Tudor's ships were sailing to Brazil and Bombay, India.

Ice-chilled drinks became a staple of life in southern states. (Even today, Americans are far more likely to enjoy ice with their beverages than Europeans, a distant legacy of Tudor's ambition.) By 1850, Tudor's success had inspired countless imitators, and more than a hundred thousand tons of Boston ice were shipped around the world in a single year. By 1860, two out of three New York homes had daily deliveries of ice. When Tudor died in 1864, he had amassed a fortune worth more than two hundred million in today's dollars.

Chicago prospered because of its railroads and slaughterhouses. But it is just as true to say that it was a city built by ice.

Cold Meat to Go

Ice-powered refrigeration changed the map of America, nowhere more so than in the transformation of Chicago. The city's initial burst of growth had come after canals and rail lines connected it to both the Gulf of Mexico and the cities of the eastern seaboard. Chicago's fortuitous location as a transportation hub—created both by nature and some of the most ambitious engineering of the nineteenth century—enabled wheat to flow from the bountiful Midwest to the heavily populated Northeast. But meat couldn't make the same journey without spoiling. In the mid-1800s, Chicago developed a large trade in preserved pork. Hogs were slaughtered in the stockyards on the outskirts of the city and salted and stored in barrels to be sent east. But fresh beef remained largely a local delicacy.

As the nineteenth century progressed, supply and demand between the hungry cities of the Northeast and the cattle ranches in the Midwest became imbalanced. Waves of immigration swelled the populations of New York, Philadelphia, and other eastern urban centers in the 1840s and 1850s. Local beef suppliers could not keep up with the surging demand for meat. Meanwhile, the conquest of the Great Plains had enabled ranchers to breed massive herds of cattle, but there were far fewer people around to feed. You could ship live cattle by train to the eastern states to be slaughtered locally, but transporting entire cows was expensive, and the animals often became malnourished or even injured en route. Almost half would be inedible by the time they arrived in New York or Boston.

Ice helped reset the balance and bring meat to more mouths.

In 1868, the Chicago pork magnate Benjamin Hutchinson built an innovative new packing plant. It featured cooling rooms full of ice, so that pork could be chilled, packed, and shipped all year round. Hutchinson's cooling rooms inspired other entrepreneurs to integrate ice-cooled facilities into the meatpacking trade. Some companies began transporting beef back east in open-air railcars during winter, relying on the outside temperature to keep the steaks cold.

In 1878, Gustavus Franklin Swift hired an engineer to build an advanced refrigerator train car; it was designed from the ground up to transport beef to the eastern seaboard year round. Ice was placed in bins above the meat; at stops along the train route, workers swapped in new blocks of ice from overhead, without disturbing the meat below. Breakthroughs in railway refrigeration led to new possibilities for refrigerator ships. Chicago's meat business went global.

The explosive success of the meat trade transformed the landscape of the American plains. Vast, shimmering grasslands were replaced by industrial feedlots. In Chicago's stockyards, fourteen million animals were slaughtered in an average year. The route from grim feedlot to bloody stockyard assembly line to long, cold refrigerated trains led to the creation of the industrial food complex.

Ice made a new kind of food network imaginable. Yet even in the middle of the nineteenth century, when coal-fired factories, railroads, telegraphs, and other innovations were rapidly changing the way people lived and worked, the ice business was still entirely based on old technology: cutting chunks of frozen water out of a river, lake, or pond. A century into the Industrial Revolution, artificial cold was still a fantasy.

The commercial demand for ice—all those millions of dollars flowing upstream from the tropics to the ice barons of New England—sent a signal out across the world that there was money to be made from cold, which inevitably sent some inventive minds off in search of the next logical step. And once again, another chapter of the story of cold begins somewhere hot, this time featuring a new character: the mosquito.

A New Kind of Cold

In you live alongside a swamp in a subtropical climate, you're living alongside mosquitoes. Lots of them. And where there are mosquitoes, there can be malaria.

In 1842, in a modest hospital in Apalachicola, Florida, Dr. John Gorrie was desperate for a way to help his patients who had been bitten by mosquitoes and were burning up with malarial fever. He suspended blocks of ice from the hospital ceiling, which cooled the air and thereby the patients. But Gorrie's clever solution was undermined by another natural element in Florida: hurricanes. When these storms caused a string of shipwrecks and delayed ice shipments, Gorrie's supply of ice ran out. The young doctor began mulling over a more radical solution for his hospital: making his own ice.

Dr. John Gorrie included a model of his ice-making machine with a patent request in 1851. But natural ice was abundant and cheap, and Gorrie got nowhere with his idea as a businessman. He died penniless, having failed to sell a single machine.

Up until the middle of the nineteenth century, the smartest mind in the world couldn't have invented a refrigerator; the necessary building blocks just weren't there yet. But by 1850, the pieces had come together. Engineers learned more about how heat and energy are converted from the steam engine. Tools for measuring heat and weight with increased precision were developed, along with standardized scales such as Celsius and Fahrenheit.

This information was all available to Gorrie, which helped him form the new connections needed to build a refrigeration machine. In Gorrie's design, energy from a pump compressed air. This compression heated the air, which was then run through pipes cooled with water. When the compressed air expanded, it pulled heat from its environment, and in turn, that heat extraction cooled the surrounding air. It could even be used to create ice. Gorrie applied for a patent for his machine, which he rightly predicted would "better serve mankind. . . . Fruits, vegetables, and meats will be preserved in transit by my refrigeration system and thereby enjoyed by all!"

The good doctor was onto something, but he was not alone. Suddenly, the idea of artificial cold was "in the air," and people around the world who had independently hit on the same essential concept filed patents. Existing scientific knowledge, combined with the fortunes being made in the ice trade, made artificial cold ripe for invention.

One of those simultaneous inventors was the French engineer Ferdinand Carré, whose design for a refrigeration machine followed the same basic principles as Gorrie's. Carré built his prototype in Paris, but its success was due to events unfolding in the United States.

After the Civil War broke out in 1861, the Union navy blockaded the South's ships to prevent trading and cripple the Confederate economy. No ships meant no ice. To help combat the resulting ice famine in the American South, Carré's machines were smuggled all the way from France to Georgia, Louisiana, and Texas. Along the way, a network of innovators tinkered

This print shows a fleet running the blockade of the Mississippi at Vicksburg, April 16, 1863. As the Civil War raged, blockade runners smuggled more than gunpowder or weapons. Sometimes they carried more novel goods, like Carré's ice-making machine.

with Carré's machines, improving their efficiency. A handful of industrialized commercial ice plants opened. By 1870, five years after the war had ended, the southern states were making more artificial ice than anywhere else in the world.

Artificial refrigeration exploded and became a huge industry. Cities, freed from the limits of their surrounding resources, experienced rapid growth. With a richer variety of food available, many people became healthier and better nourished. Artificial refrigeration was shaping a new America.

Frozen in a Flash

In the winter of 1916, an eccentric naturalist and entrepreneur moved his young family to the remote tundra of Labrador in northern Canada. Clarence Birdseye had spent several winters there on his own, starting a fur company and writing scientific reports for the U.S. government.

Winter temperatures in Labrador regularly hit 30 degrees below Fahrenheit. The bleak climate meant that your winter menu was typically frozen fish or "brewis," a local dish of salted cod and hardtack. (Hardtack is rock-solid bread, which was boiled and garnished with "scrunchions," small, fried chunks of salted pork fat.) Any meat or produce from warmer months that had been frozen was always mushy and tasteless when thawed out.

Birdseye took up ice fishing with some of the local Inuit, carving holes in frozen lakes and casting a line for trout. In the subzero air temperatures, a fish pulled out of the lake froze solid in a matter of seconds. But later, when the trout was thawed out, cooked, and put on the table, it tasted far better than the usual grub. Birdseye became obsessed with finding out why.

At first, he assumed the fish tasted fresher simply because it had been caught more recently. But the deeper Birdseye got into his studies, the more he suspected there was some other factor at work. For starters, ice-fished catch retained its flavor for months, unlike other food preserved through

ICE UP TO 40 CENTS AND A FAMINE IN SIGHT

New York's Visible Supply Less Than a Third of Last Year's.

EVEN MAINE'S CROP FAILED

Despite the Advance in Prices the Ice Company Men Say They'll Lose Money This Summer.

A 1906 New York Times *headline* (above). *In less than a century, ice had gone from a curiosity to a luxury to a necessity. An unusually warm winter could set off a frenzy of speculation about an "ice famine."*

Clarence Birdseye in Labrador, Canada, 1912. The naturalist was an adventurous eater, fascinated with the cuisines of different cultures. In his journals, he recorded eating everything from rattlesnake to skunk.

A worker in New Jersey packs vegetables for flash freezing on the Birdseye production line, 1942.

freezing. He began experimenting with vegetables and discovered that those frozen in the depths of winter tasted better than produce frozen in late fall or early spring. (It was still cold enough to freeze food in Labrador in March or November, though not nearly as cold as January.) Birdseye analyzed the food under a microscope. He noticed a striking difference in the ice crystals that formed during the freezing process: the flavorless frozen vegetables had significantly larger crystals that seemed to break down the molecular structure of the food itself.

Eventually, Birdseye figured it out: it wasn't just temperature; it was speed. A slow freeze allowed the hydrogen bonds of ice to form larger crystalline shapes. But a freeze that happened in seconds—"flash freezing," as we now call it—generated much smaller crystals that did less damage to the food. This explained why food frozen in January tasted better than food frozen in March: the relatively warmer early spring air meant that the food took longer to freeze. Inuit fishermen had been savoring the benefits of flash freezing for centuries by pulling live fish out of the water into shockingly cold air.

After his Labrador adventure, Birdseye and his family returned to his original home in New York. Clarence took a job with the Fisheries Association and saw firsthand the foul containers of cod-fishing trawlers in New York harbors, from which came some of the bottom-of-the-barrel stuff that was the frozen-food business in the early twentieth century. He continued taking notes on his experiments with cold and realized that with artificial refrigeration becoming increasingly commonplace, the market for frozen food could be big. But it took nearly ten years for his slow hunch to turn into something commercially viable.

By the early 1920s, Birdseye had developed a flash-freezing process for stacked cartons of fish frozen at minus 40 degrees Fahrenheit. Inspired by Henry Ford's new industrial assembly line at his Model T factory, Birdseye created a "double-belt freezer" process that ran along an efficient production line. He found that just about anything he froze with this method—fruit, meat, vegetables—was remarkably fresh after thawing. He formed a company called General Seafood and started production using the double-belt freezer.

Flash-frozen products extended the reach of the food network across space and time. Fish caught in the North Atlantic could be eaten in Denver or Dallas. Produce harvested in summer could be consumed months later; you could eat frozen peas in January and not have to wait five months for fresh ones.

Birdseye's experiments were so promising that in 1929, just months before the stock market crash that brought on the Great Depression, General Seafood was acquired by another company, making him a multimillionaire. Like every big idea, Birdseye's breakthrough was not a single insight, but a *network* of other ideas, packaged together in a new way.

Life Becomes Chill

Eventually, an entire economy of cold developed. Inventors like Frederick McKinley Jones made important contributions to that economy. In the 1930s, Jones designed a small, durable refrigerated unit that mounted on a truck and kept its contents chilled. After World War II, he developed refrigerated containers that could be moved from train to ship to truck, perfecting America's food distribution.

By the 1950s, Americans had adopted a lifestyle that was profoundly shaped by artificial cold, buying frozen dinners from the refrigerated aisles of their local supermarkets and stacking them up in the deep freeze of their new Frigidaires, which featured the latest in ice-making technology. But in many places in the United States, while refrigerators kept the food cold, people themselves were still too hot.

The first "apparatus for treating air"—what we now call an air conditioner—had been dreamed up by a young engineer named Willis Carrier in 1902. His invention is a classic story of accidental discovery. When he was a young engineer, Carrier had been hired by a Brooklyn printing company with a problem: ink smeared in the printing process during humid summer

Frederick McKinley Jones. Although he lived at a time when African American inventors were rarely recognized or given opportunities, Jones filed more than forty cooling patents. He revolutionized mobile refrigeration technology and co-founded the Thermo-King refrigeration company in 1938.

A gift in a million...for a wife in a million!

8-cu-ft model (NH-8), illustrated. Also available in 10-cu-ft size. Features include special butter conditioner in door . . . ample bottle space with room for tall bottles . . . sliding shelves . . . *two* deep drawers for fruits and vegetables (can be stacked to make extra room for bulky items). Freezer compartment has 3 ice trays and covered dessert pan.

General Electric 1949 Two-door Refrigerator-Home Freezer Combination

This year—if you want to make your wife the happiest woman in the world—let your major present be a new General Electric Refrigerator-Home Freezer Combination.

You might not appreciate all that it means to have this most advanced refrigerator.

But you can be sure your wife will! She'll know you're giving your family years and years of better living—greater kitchen convenience—tastier foods on the table—and new economies in buying and keeping foods.

She'll fall in love with that big, separate home freezer compartment, with its own separate door. For it freezes foods and ice cubes quickly . . . maintains zero temperature at all times! The 10-cubic-foot model holds up to 70 pounds of frozen foods.

And she'll thrill over the moisture-conditioned refrigerator compartment that gives as much refrigerated fresh-food storage space as in ordinary 8- and 9-cubic-foot refrigerators!

It never needs defrosting . . . no need to cover dishes.

And she'll know, of course, that the General Electric trademark means utmost dependability . . . dependability based on an unexcelled record for year-in, year-out performance.

We can't begin to tell you here the story of this most wonderful of gifts for the home.

So why not do this: Take your wife to the nearest General Electric retailer. Let him give you a demonstration of the General Electric Refrigerator-Home Freezer Combination.

Then—later on—when your wife gets through talking about how much she'd like one of those great refrigerators, just say quietly: "I'm giving you one for Christmas, darling!"

General Electric Company, Bridgeport 2, Connecticut.

More than 1,700,000 General Electric Refrigerators in service ten years or longer.

GENERAL ⊕ ELECTRIC

months. The invention Carrier came up with removed the humidity from the printing room; it also chilled the air. Carrier noticed that everyone suddenly wanted to have lunch next to the printing presses!

The enterprising engineer started designing contraptions to regulate the humidity and temperature in an interior space. Within a few years, Carrier formed a company that focused on industrial uses for his technology. But he was convinced that air-conditioning should also benefit the public. And where could you often find the public? At the movies—except in summer. Nobody would pay to sit in a dark, stuffy, hot room with several hundred other sweaty bodies, no matter which glamorous Hollywood stars were on the marquee. Naturally, movie studios wanted to change that. Carrier wanted to show that his technology could.

On Memorial Day weekend of 1925, Carrier debuted his experimental air-conditioning system at the Rivoli, Paramount Pictures' new flagship movie theater in Manhattan. He had persuaded Adolph Zukor, the legendary chief of Paramount, that there was money to be made by investing in central air-conditioning for his theaters. Zukor himself had hunkered down inconspicuously in a balcony seat. Moviegoers throughout the theater were furiously waving hand fans to keep cool before the movie started. Carrier and his team had some technical difficulties at first, but then got the system running. The room began to cool, people stopped fanning themselves, and Zukor saw the future: "Yes, the people are going to like it."

Between 1945 and 1949, Americans bought twenty million refrigerators to help store the 300,000 tons of frozen food that were being sold every year. Freezer trucks, refrigerated warehouses, supermarkets with freezer units, and an electrical grid powering new suburban homes all contributed toward putting an electric refrigerator in every kitchen. This 1949 ad shows who would be responsible for all those appliances and the frozen food they held: women, who were expected to be in the kitchen, too.

Willis H. Carrier, circa 1950. Carrier's work at a printing press led to the first modern air-conditioning system.

For the next twenty-five years or so, most Americans experienced air-conditioning only in large commercial spaces such as movie theaters, department stores, hotels, or office buildings. Willis Carrier knew that air-conditioning was headed for homes, too, but the current machines were simply too large and expensive for residences.

Carrier's vision of Americans chilling out at home was postponed by the outbreak of World War II, but by the late 1940s, the first in-window portable air-conditioning units appeared on the market. Within five years, Americans were installing more than a million units a year. A machine that had once been larger than a flatbed truck now sat snugly in living room and bedroom windows across the country.

Carrier's invention circulated more than just molecules of air and water. It ended up circulating *people* as well. By the 1960s, the population of the

Sun Belt—the region stretching southeast to southwest along the lower part of the United States—was exploding. Thanks to domestic air-conditioning, immigrants from colder states packed up and headed out in droves to settle in places where the tropical humidity or blazing desert climate once would have discouraged new construction and real estate signs. The population of Tucson, Arizona, grew 400 percent, rocketing from 45,000 to 210,000 inhabitants in just ten years; Houston, Texas, expanded from 600,000 to 940,000 in the same decade. Just fifty years after Carrier's air-conditioning debuted at the Rivoli Theatre, Florida's population had jumped from less than one million to more than twenty million, thanks in large part to air-conditioned homes.

In the 1930s, eighty million Americans—65 percent of the entire population—went to the movies every week. Carrier's invention kept them coming over the hot summer months to enjoy a film and be "cooled by refrigeration."

Temperatures in Dubai in the United Arab Emirates can reach over 100 degrees Fahrenheit——but this indoor ski slope is kept at a chilly 28 degrees by the combination of insulation and artificial refrigeration.

This large-scale migration changed the political map of America. Swelling populations in Florida, Texas, and Southern California shifted the electoral college (the process established in the U.S. Constitution that determines the election of the president). Warm-climate states gained twenty-nine electoral college votes between 1940 and 1980, while the colder states of the Northeast and Rust Belt (the declining, once heavily industrial section of the country that stretches from the Great Lakes to the upper Midwest states) lost thirty-one. In the first half of the twentieth century, only two presidents or vice presidents hailed from Sun Belt states. But starting in 1952, every single winning presidential ticket contained a Sun Belt candidate, until Barack Obama and Joe Biden broke the streak in 2008. Politicians and political strategists from both the Republican and Democratic parties remain keenly aware of how important it is for their candidates to connect with voters from Sun Belt states.

What happened in the United States is now happening on a planetary scale, too. The world's fastest-growing megacities are predominantly in tropical climates: Chennai, Bangkok, Manila, Jakarta, Karachi, Lagos, Dubai, Rio de Janeiro. More than a billion new residents are predicted to be living in these cities by 2025. It goes without saying that many people in these places don't have air-conditioning in their homes, at least not yet. It's also an open question whether these cities are environmentally sustainable in the long run, particularly those in desert climates. But the ability to control temperature and humidity allowed these urban centers to achieve megacity status. (It's no accident that the world's largest cities—London, Paris, New York, Tokyo—were almost exclusively in temperate climates until the second half of the twentieth century.) What we're seeing now is the first mass migration of humans made possible by a home appliance, and the long-zoom history of big thinkers who were hot on the possibilities of cold.

Sound

———————————————————————▶

You can put on earphones and listen to anything, right? Music, podcasts, audiobooks: you're living in a modern sonic world. But did you ever think about listening to the past—as in, the very distant, cave-dwelling past?

Women on the production line at a vinyl record factory in West London, England, assemble the Beatles' latest album Rubber Soul, *circa 1965.*

In the early 1990s, an incredible collection of ancient paintings was discovered in a cave complex in Arcy-sur-Cure, France. The rock walls were covered with more than a hundred images of bison, woolly mammoths, birds, fish, even the haunting imprint of a child's hand. Archeological findings suggest that Neanderthals and early modern humans used these caves for shelter and ceremony for tens of thousands of years. Radiometric dating, which is a process that measures the age of certain geological or organic matter, determined that the images were thirty thousand years old.

Cave paintings are usually cited as evidence of a deep human need to represent the world in images. But in recent years, a new theory has emerged about the primitive use of caves, one focused not on the pictures in these underground passages but rather on the *sounds*.

Rocking the Cave

A few years after the paintings in Arcy-sur-Cure were discovered, Iegor Reznikoff, a music professor from the University of Paris, began studying the echoes and reverberations created in different parts of the cave complex. The Neanderthal art was clustered in specific areas, with some of the most ornate and dense images appearing more than two-thirds of a mile (one kilometer) deep. Reznikoff determined that the paintings were consistently placed at the most acoustically interesting parts of the cave, where reverberation was the most profound: if you make a loud sound while standing beneath the images of Paleolithic animals at the far end of the Arcy-sur-Cure caves, you hear seven distinct echoes of your voice. The reverberation takes almost five seconds to die down after your vocal cords stop vibrating.

Reznikoff's theory is that Neanderthal communities gathered for rituals beside the images they had painted. They chanted or sang, using the reverberations of the cave to magically widen the sounds. If the professor is correct, those early humans were experimenting with a primitive form of audio engineering, amplifying that most intoxicating of sounds: the human voice.

The drive to enhance—and, ultimately, reproduce—the human voice would in time pave the way for a series of breakthroughs in communications, computation, politics, and the arts.

The technologies of the voice did not arrive in full force until the late nineteenth century. When they did, they changed just about everything. But they didn't begin with amplification. The first great breakthrough in our obsession with the human voice arrived in the simple act of writing it down.

Shorthand for Sound Waves

Recording the human voice became possible only after two key developments, one from physics and the other from anatomy. From about 1500 on, scientists worked under the assumption that sound traveled in invisible waves. By the eighteenth century, detailed books of anatomy had mapped the basic structure of the human ear and documented the way sound waves were funneled through the auditory canal and triggered vibrations in the eardrum. In the 1850s, a Parisian printer named Édouard-Léon Scott de Martinville stumbled across one of these anatomy books, triggering a hobbyist's interest in the biology and physics of sound.

Scott had also been a student of shorthand writing, or stenography, which uses abbreviations and symbols for words and phrases so that you can take notes very quickly while someone is talking. At the time, stenography was the most advanced form of voice-recording technology in existence; no other system captured the spoken word with as much accuracy and speed. When Scott looked at the detailed illustrations of the inner ear, a slow hunch took shape. What if the process of transcribing the human voice could be automated? Instead of a human writing down words, what if a machine could write sound waves?

Scott got to work on a contraption that funneled sound waves through a hornlike apparatus that ended with a parchment membrane. The waves triggered vibrations in the parchment, which were then transmitted to a stylus made from the stiff hair of a pig. The stylus etched the waves on a page darkened by lampblack, the carbon soot collected from oil lamps. He called his invention the "phonautograph": the self-writing of sound.

Scott had made a critical conceptual leap—that sound waves could be pulled out of the air and etched onto a recording medium—more than a decade before other inventors and scientists got around to it. He had invented the first sound-recording device in history. But there was one big drawback to his phonautograph: *no playback.*

Édouard-Léon Scott de Martinville patented his phonoautograph in March 1857, twenty years before Thomas Edison invented the phonograph.

It seems obvious to us now that a sound recorder should also include a feature that enables you to *hear* the recording. But in the 1850s, this was not an intuitive idea. The issue wasn't so much that Scott forgot or failed to make playback work; it was that the idea never even occurred to him. He thought of recording audio through the metaphor of stenography, only with sound waves instead of words. With stenography, the information noted can be read by someone who understands the code. Scott thought the same thing would happen with his phonautograph. The machine would record waveforms of someone speaking, and people would learn to "read" those squiggles the way they had learned to read shorthand.

In a sense, Scott wasn't trying to invent an audio-recording device at all. He was trying to invent the ultimate transcription service—only you had to learn a whole new language in order to read the transcript. Unfortunately, our neural toolkit doesn't seem to include the capacity for reading sound waves by sight. If we were going to decode recorded audio, we needed to convert it back to sound so we could do our decoding via the eardrum, not the retina.

Can You Hear Me Now?

In 1872, another inventor modified Scott's original design to include an actual ear from a cadaver in order to understand the acoustics better. Through his tinkering, he hit upon a method for both capturing *and* transmitting sound, so that the human voice could be sent along already-existing telegraph wires. That inventor was Alexander Graham Bell, who became famous for his telephone.

Before the telephone, people usually communicated across distances via the postal service. Bell's invention revolutionized that. The impact of the telephone was far-reaching and brought the world closer together, though there were few lines connecting us all until recently. The first transatlantic line that enabled ordinary citizens to call between North America and Europe was laid just over sixty years ago, in 1956, and the system could only handle twenty-four simultaneous calls. That was the total bandwidth for a combined population of several hundred million people on the two continents: two dozen transatlantic conversations at a time.

The Invisible Becomes Audible

Perhaps the most significant long-zoom legacy of the telephone lay in the groundbreaking organization that grew out of it: Bell Labs, a research-and-development group originally formed by Alexander Graham Bell in the 1880s. Bell Labs played a critical role in developing almost every major technology of the twentieth century. Radios, vacuum tubes, transistors, televisions, solar cells, coaxial cables, laser beams, microprocessors, computers, cell phones, fiber optics—all these essential tools of modern life descend from ideas originally generated at Bell Labs. Not for nothing was it known as "the idea factory."

With the telephone, we crossed a crucial threshold in the history of technology: a component of the physical world, in this case sound, was represented in electrical energy in a direct way. Someone spoke into a receiver, generating sound waves that became pulses of electricity that became sound waves again on the other end. Once those sound waves became electric, they could travel vast distances at astonishing speeds. But they were not infallible. Traveling from city to city over copper wires, the electric pulses were vulnerable to decay, signal loss, and noise. Amplifiers, as we will see, helped combat the problem, boosting signals as they traveled down the line. But the ultimate goal was a pure signal, some kind of perfect representation of the voice that wouldn't degrade as it wound its way through the telephone

Alexander Graham Bell (center) at the opening of the long-distance line from New York to Chicago, October 18, 1892. Bell thought the telephone would be used for sharing live music. An orchestra would perform on one end of the line, and listeners would enjoy the sound through the telephone speaker on the other. Likewise, when Edison invented the phonograph, he imagined it would be used as a means of communication; people would mail audio letters recorded on the phonograph's wax scroll. The two legendary inventors had things exactly reversed.

network. Interestingly, the path that ultimately led to that goal began with a different objective: not keeping our voices pure, but keeping them *secret*.

Who's Calling? . . . Who Wants to Know?

During World War II, the legendary British mathematician Alan Turing and A. B. Clark of Bell Labs collaborated on a secure communications line. Code-named SIGSALY, the system recorded a sound wave twenty thousand times a second, but instead of converting the recording into an electrical signal or a groove in a wax cylinder, it turned the information into numbers, digitally encoding it in the binary language of zeros and ones, the language that all modern computers use.

The power of digital programming had first been glimpsed more than a century before by the Victorian inventor Charles Babbage and his collaborator Ada Lovelace. Babbage built a machine in the 1830s called the Analytic Engine that many consider to be the first programmable computer, and Lovelace—who was a brilliant mathematician in an era when young women were actively discouraged from studying male-dominated fields like math and science—wrote the first lines of code for it. (Today, many people in the tech community celebrate Ada Lovelace's birthday on December 10th as a reminder that the first computer programmer was a woman.)

Lovelace also made another important contribution to the digital world, one that connects directly to SIGSALY and its legacy. In the notes that accompanied her early software programs, she suggested a radical idea: that computers might be good for creative work, not just crunching numbers. She imagined a future where "the Engine might compose elaborate and scientific

An 1836 portrait of Ada Lovelace, painted by British artist Margaret Sarah Carpenter.

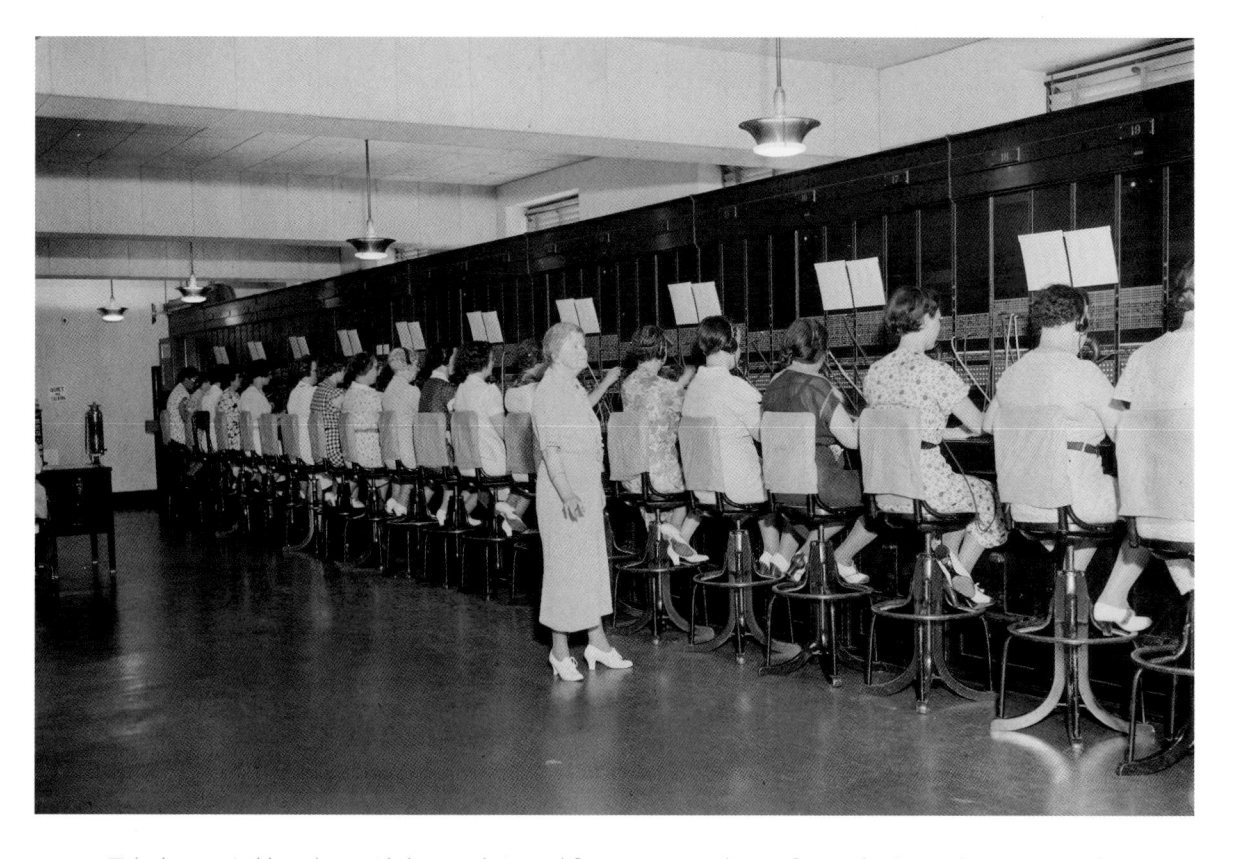

Telephone switchboards provided an early inroad for women into the "professional" class. The U.S. Capitol building got a switchboard in 1898, operated by one woman, Harriott Daley (shown here standing and still on the job in 1937). By the 1940s, fifty operators, known as "Hello Girls," were answering calls in the Capitol.

pieces of music of any degree of complexity or extent." The idea that what seemed like a giant calculator could be used to create music seemed preposterous to most of Lovelace's peers. But she turned out to be right.

The key step that turned Lovelace's visionary idea into reality was thinking about sound in terms of digital code. With SIGSALY, this meant taking the sound of a military communication and transforming it into a series of zeros and ones that described the sound wave so that it could be converted back into intelligible speech on the receiving end. Initially, this was all in the service of keeping those messages secret. It was much easier to transmit these digital samples securely. Anyone looking for a traditional analog signal would just hear a blast of digital noise. (SIGSALY was nicknamed "Green Hornet" because the noise sounded like a buzzing insect in the theme music of a popular radio show.) While the Germans intercepted and recorded many hours of SIGSALY transmissions, they were never able to interpret them.

SIGSALY went into operation on July 15, 1943, but the truly disruptive force that it unleashed came from another of its powerful capabilities: the fact that digital copies could be perfectly duplicated. With the right equipment, digital samples of sound could be transmitted and copied with true fidelity, or they could be manipulated to make new kinds of music. Lovelace's brilliant prediction proved true with this digital revolution in sound. Once we began making digital copies of songs, movies, and television shows, as well as using computers and synthesizers to record and create new forms of electronic music, the entertainment business was transformed. So many of the changes in modern media—the reinvention of the music business that began with file-sharing services, the market impact of iTunes, the rise of streaming media, the pulsing sounds of most Top 40 pop songs today—can be traced back to the digital buzz of the Green Hornet and the innovative ideas of Ada Lovelace.

Telegraph + Telephone = *Radio?*

SIGSALY's digital samples traveled courtesy of another communications breakthrough that Bell Labs helped create: radio. Interestingly, while radio would eventually become a medium saturated with the sound of people talking or singing, it did not begin that way. The first functioning radio transmissions—produced by Italian electrical engineer Guglielmo Marconi and a number of other more-or-less-simultaneous inventors in the last decades of the nineteenth century—were almost exclusively devoted to sending Morse code. (Marconi called his invention "wireless telegraphy.") But

once information began flowing through the airwaves, it was not long before the tinkerers and research labs began exploring how to make spoken word and song part of the mix.

One of those tinkerers was Lee de Forest, a brilliant and erratic inventor. Working out of his home lab in Chicago, de Forest dreamed of combining Marconi's wireless telegraph with Bell's telephone. He began a series of experiments with a spark-gap transmitter, a device that created a bright, monotone pulse of electromagnetic energy that can be detected by antennae miles away, perfect for sending Morse code. One night, while de Forest was triggering a series of pulses, he noticed something strange happening across the room: every time he created a spark, the flame in his gas lamp turned white and increased in size. Somehow, de Forest thought, the electromagnetic pulse was intensifying the flame. That flickering gaslight planted a seed in his head: could a gas be used to amplify weak radio reception? Maybe even making it strong enough to carry the signal of spoken words and not just the staccato pulses of Morse code?

After a few years of trial and error, de Forest settled on a gas-filled bulb containing three precisely configured electrodes designed to amplify incoming wireless signals. He called it the Audion. As a transmission device for the spoken word, the Audion was just powerful enough to transmit intelligible signals. In 1910, de Forest used an Audion-equipped radio device to make the first-ever ship-to-shore broadcast of the human voice. But de Forest had much more ambitious plans for his device. He imagined a world in which his wireless technology was used not just for military and business communications but also for mass enjoyment—and in particular, to make his great passion, opera, available to everyone.

Lee de Forest at the end of the 1920s.

On January 13, 1910, during a performance of *Tosca* by New York's Metropolitan Opera, de Forest hooked up a telephone microphone in the opera hall to a transmitter on the roof to create the first live public radio broadcast. He invited hordes of reporters and VIPs to listen on his radio receivers dotted around the city. The signal strength was terrible; the *New York Times* declared the whole adventure "a disaster." De Forest was even sued by the U.S. attorney for fraud, accused of overselling the value of the Audion in wireless technology. Needing cash to pay his legal bills, de Forest sold the Audion patent at a bargain price to American Telephone and Telegraph (AT&T), a Bell company.

A Jazzy Breakthrough

When the researchers at Bell Labs began investigating the Audion, they discovered something extraordinary: Lee de Forest's slow hunch was flat-out wrong. The increase in the gas flame had nothing to do with electromagnetic radiation. The flare was caused by sound waves from the loud noise of the spark in de Forest's experiment. Gas didn't detect and amplify a radio signal at all.

Over the next decade, engineers at Bell Labs and elsewhere modified de Forest's basic three-electrode design. They removed the gas from the bulb, making it a perfect vacuum and transforming it into both a transmitter *and* a receiver. The result was the vacuum tube, the first great breakthrough of the electronics revolution. Vacuum tubes boost the electrical signal of just about any technology. Television, radar, sound recording, guitar amplifiers, X-rays, microwave ovens, early digital computers—all would rely on vacuum tubes. But the first mainstream technology to bring the vacuum tube into the home was radio.

In the 1940s, singers like Ella Fitzgerald, shown here, appeared on stage and on live radio shows, spreading the jazz sound.

Radio began as a two-way medium, a practice that continues to this day as ham radio: individual hobbyists talking to one another over the airwaves, occasionally eavesdropping on other conversations. By the early 1920s, the broadcast model had evolved. Professional radio stations began delivering packaged news and entertainment to audiences who listened on receivers in their homes.

And something entirely unexpected happened.

The existence of a mass medium for sound unleashed a new kind of music on the United States, a music that had until then belonged almost exclusively to New Orleans, to the river towns of the American South, and to African American neighborhoods in New York and Chicago. Almost overnight, radio made jazz a national phenomenon. It created new African American celebrities, like Duke Ellington, Louis Armstrong, and Ella Fitzgerald. It was a profound breakthrough: for the first time, white America welcomed African American culture into its living room, albeit through the speakers of a radio.

The popularity of jazz changed more than our musical tastes. The birth of the civil rights movement was intimately bound up in the spread of jazz music throughout the United States. It was, for many Americans, the first cultural common ground between black and white America that had been largely created by African Americans, which in itself was a blow to segregation. Martin Luther King Jr. made the connection explicit in remarks he delivered at the Berlin Jazz Festival in 1964: "Much of the power of our Freedom Movement in the United States has come from this music."

Can You Hear Me in the Back?

Like many political figures of the twentieth century, King was indebted to the vacuum tube for another reason. Shortly after de Forest and Bell Labs began using vacuum tubes to enable radio broadcasts, the technology was

Dr. Martin Luther King Jr. speaks at an event on August 28, 1963. Microphones allowed Dr. King's historic "I Have a Dream" speech to be heard by more than 250,000 people at the March on Washington for Jobs and Freedom.

enlisted to amplify the human voice in more immediate settings by powering amplifiers attached to microphones. This allowed people to speak or sing to massive crowds for the first time in history. We were no longer dependent on the reverberations of caves or cathedrals or opera houses to make our voices louder. Now electricity could do the work of echoes, but a thousand times more powerfully.

Amplification created an entirely new kind of political event: mass rallies oriented around individual speakers. Before tube amplifiers, the limits of our vocal cords made it difficult to speak to more than a thousand people at a time. But a microphone attached to multiple speakers extended the range of earshot by several orders of magnitude.

Tube amplification enabled the musical equivalent of political rallies as well: Woodstock, Live Aid, the Beatles playing Shea Stadium. But the idiosyncrasies of vacuum tube technology also had a more subtle effect on twentieth-century music, making it loud *and* noisy. Starting in the 1950s, guitarists playing through tube amplifiers noticed that they could make an intriguing new kind of sound. By overdriving the amp, they generated a crunchy layer of noise on top of the notes made by strumming the strings of the guitar itself. This was, technically speaking, the sound of the amplifier malfunctioning, distorting the sound it had been designed to reproduce. A handful of early rock 'n' roll recordings feature some distortion on the guitar tracks, but the distorted sound we now call a "fuzz tone" didn't really take off until the sixties, when commercial distortion boxes came on the market and musicians like Keith Richards of the Rolling Stones played now-legendary fuzz riffs like the opening notes of "(I Can't Get No) Satisfaction."

A similar pattern developed with feedback, the swirling, screeching sound that occurs when amplified speakers and microphones share the same physical space. In the late 1960s, musicians such as Jimi Hendrix created a new sound that drew upon the vibration of the guitar strings, the microphone-like pickups on the guitar itself, and the speakers, building on the complex and unpredictable interactions among those three technologies.

Deep-"See" Listening

Sometimes innovations come from using new technologies in unexpected ways. From the start, the story of sound technology had always been about extending the range and intensity of our voices and our ears. But the most surprising twist of all came just a century ago, when humans first realized that sound could be harnessed for something else: to help us see.

Since ancient times, lighthouses have been built to signal the presence of dangerous shorelines to sailors. But lighthouses perform poorly precisely when they are needed the most: in stormy weather, when the light they transmit is obscured by fog and rain. Many lighthouses employed warning bells as an additional signal, but they could be drowned out by a roaring sea. Still, sound waves turned out to have an intriguing physical property. They travel four times faster underwater than they do through the air, and underwater sound waves are largely undisturbed by any sonic chaos above sea level.

In 1901, a Boston-based firm called the Submarine Signal Company (SSC) began manufacturing a system of communications tools that exploited this property of aquatic sound waves. In their system, underwater bells chimed at regular intervals and were picked up by "hydrophones," microphones specially designed for underwater reception. The SSC established more than a hundred stations around the world at particularly treacherous harbors or channels. It was an ingenious system, but it had its limits. To begin with, it only worked in places where the SSC had installed warning

Reginald Fessenden lowering his oscillator while on board a ship, circa 1915.

bells. And it was entirely useless at detecting less predictable dangers such as other ships or icebergs.

The threat posed by icebergs to maritime travel was headline news on April 15, 1912, when the luxury liner *Titanic* hit one and sank in the North Atlantic. Just a few days before this catastrophe, the Canadian inventor Reginald Fessenden had visited the SSC office to check out the latest underwater signaling technologies. Fessenden was a pioneer of wireless radio and was responsible for both the first radio transmission of human speech and the first transatlantic two-way radio transmission of Morse code. Due to his expertise, the SSC had asked him to help them design a better hydrophone system that would filter out the background noise of underwater acoustics. When news of the fatal shipwreck broke, just four days after his visit to the SSC, Fessenden was as shocked as the rest of the world. Unlike the rest of the world, though, he had an idea about how to prevent such tragedies in the future.

Ahoy! Iceberg! Or Spy Sub! Or Fish!

Fessenden invented a device that generated its own sounds and then listened to the echoes created as those sounds bounced off objects in the water, much as dolphins use echolocation to navigate their way around the ocean. His Fessenden

The wreckage of the Titanic. *In 1985 a team of American and French researchers used sonar——first dreamed up by Fessenden when the* Titanic *sank——to locate the ship in the Atlantic twelve thousand feet below the surface.*

oscillator could be used onboard ships to detect objects up to two miles away. He tuned it to ignore all the background noise of the aquatic environment. It was both a system for sending and receiving telegraphy and the world's first functional sonar device.

Just a year after Fessenden completed his first working prototype, World War I broke out. The German U-boats (submarines) roaming the North Atlantic posed an even greater threat to maritime travel than the *Titanic*'s iceberg. But the United States was still two years away from joining the war. Executives at the Submarine Signal Company, faced with the financial risk of developing *two* revolutionary new technologies (aquatic telegraphy and sonar), decided to build and market the oscillator as a listening device only.

Fessenden tried persuading the British Royal Navy to invest in his oscillator, but his pleas were ultimately ignored. By the end of World War I in 1918, upward of ten thousand lives had been lost to the U-boats. The most valuable defensive weapon would have been a simple sound wave, bouncing off the hull of the attacker, as Fessenden had proposed. However, sonar would not become a standard component of naval warfare until World War II.

In the second half of the twentieth century, the principles of echolocation were applied to more than detecting icebergs and submarines. Fishing boats used variations of Fessenden's oscillator to detect and track their catch. Scientists used sonar to explore the last great mysteries of our oceans, revealing hidden landscapes, natural resources, and earthquake fault lines.

Fessenden's innovation had a transformative effect on medicine as well. Ultrasound devices use high-frequency sound waves, a technology similar to sonar, for live, internal views of the human body. These images provide valuable information for medical diagnosis and treatments. Ultrasounds revolutionized prenatal care, and they routinely save today's babies and their mothers from complications that would have been fatal less than a century ago.

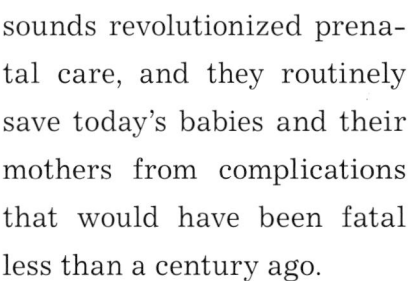

It's amazing to think about how many advances, in so many fields, owe a debt to our efforts to reproduce sound waves, starting with the sound of our own voices: new political movements, new ways to protect our ships at sea, new media platforms, new ways to keep children healthy. Sound recording has become a part of who we are.

When we launched the *Voyager* interstellar mission in 1977, one of the main objects NASA included inside the spacecraft to represent all of humanity was a gold-plated phonograph disc. It's our gift to unknown civilizations. *Voyager 1* left the solar system in 2013, becoming the first human-made object to travel into interstellar space. It will be roughly forty thousand years before it encounters another planetary system. But when it does, it will be carrying the sound of the human voice saying hello.

Astronomer Carl Sagan chose the contents of the "Golden Record" that was launched on NASA's Voyager spacecraft in 1977. The twelve-inch gold-plated copper disk contains natural sounds such as thunder and whale songs, music ranging from classical to '50s rock 'n' roll, and greetings in fifty-five different languages. Instructions and tools for playing the record are included.

Clean

In the morning when you hit the bathroom, you probably wash, brush, and flush without giving a thought to where everything goes. But look back just a century and a half and sanitation was a whole other story. Drinking water from a tap was gambling with your life. If you lived in a city, the streets outside your door were almost knee-high with refuse; pigs roamed around, devouring the waste. Life in urban America in the mid-1800s was filthy. And the filthiest place of all was Chicago.

Due to its central role as a hub for transporting wheat and meat from the Great Plains to coastal cities, Chicago experienced tremendous growth in the nineteenth century. But one thing remained

Drying laundry hangs between tenement buildings in New York City, circa 1905.

flat: its terrain. Most cities enjoy a descending grade down to the rivers or harbors they evolved around. Chicago, thanks to the legacy of a glacier's crawl thousands of years ago, is flat as an ironing board. And flat topographies don't drain—which was an issue in the middle of the nineteenth century, when gravity-based drainage was key to urban sewer systems.

Chicago also had topsoil that didn't drain well. With nowhere for the water to go, heavy summer rainstorms turned the city into a murky marshland in a matter of minutes. Chicagoans erected wood-plank roadways over the mud, but whenever a board gave way, the wet earth oozed through.

Fatal Filth

The midwestern city's rapid growth rate created housing and transportation challenges, but the biggest strain of all came from something more scatological: when almost a hundred thousand new residents move in, they generate a lot of excrement. Chicago, of course, had animal waste to deal with as well, between all the horses in the streets and the pigs and cattle awaiting slaughter in the stockyards. This filth wasn't just offensive to the eye and nose; it was deadly. Epidemics erupted regularly in the 1850s, as they did in other crowded cities in the United States and around the world.

The connection between waste and disease was not fully understood at the time. Many city authorities subscribed to the "miasma" theory. They believed that miasma, or bad air with terrible odors, spread epidemics like cholera, an often fatal intestinal disease, or dysentery, which caused dangerously severe diarrhea. The pioneering nurse and health care advocate Florence Nightingale pronounced that the "first rule of nursing" was to "keep the air within as pure as the air without." The true transmission route—invisible bacteria carried in fecal matter that polluted a water supply—would not be widely known and accepted for another decade.

Still, city leaders assumed that cleaning up would help fight disease. In 1855, a Chicago Board of Sewerage Commissioners was created to address the waste situation. They wisely hired engineer Ellis Chesbrough, whose background in railway and canal engineering turned out to be crucial to solving the problem of Chicago's flat, nonporous terrain.

Raise It Higher!

Chesbrough had a radical idea for fixing the drainage issue. Creating an artificial grade by building sewers deep underground was going to be too expensive. So Chesbrough, inspired by his railway days, decided, if you can't dig *down* to create a proper grade for drainage, why not lift the city *up*?

Ellis Chesbrough, circa 1870. Chesbrough visited many European towns, including London, Paris, Hamburg, and Amsterdam, to study their sewers.

Chesbrough's plan for building the first comprehensive urban sewer system in America was inspired by the jackscrew, a simple device that he'd seen lift multiton locomotives on and off railroad tracks. With the help of George Pullman, who would later make a fortune building railway cars, Chesbrough launched one of the most ambitious engineering projects of the nineteenth century: building by building, Chicago was lifted by an army of men with jackscrews. As the jackscrews raised the city structures inch by inch, workmen dug holes under their foundations and installed thick timbers

A NEW AMERICAN INVENTION: RAISING AN HOTEL AT CHICAGO.

to support them. Masons then scrambled to assemble a new footing under the structures. Sewer lines were installed beneath the buildings and connected to main lines running down the center of streets, which were then buried in landfill that had been dredged out of the Chicago River.

In 1860, engineers raised half a city block; almost an acre of five-story buildings, weighing an estimated thirty-five thousand tons, was lifted by more than six thousand jackscrews. By the time Chesbrough and his team finished his master plan nearly two decades later, the entire city of Chicago had been raised an average of almost ten feet. The result was the first comprehensive sewer system in any American city.

Raising the Briggs House, circa 1857. Amazingly, life went on as Chesbrough's team jacked up Chicago.

Going Subterranean

Within three decades, more than twenty cities around the country followed Chicago's lead and installed their own underground networks of sewer tunnels. Looking through the long-zoom lens of history, we see that these massive engineering projects ultimately defined the modern metropolis. They created the idea of a city as a system supported by an invisible network of subterranean services that went beyond waste treatment. The first steam train chugged through tunnels beneath London in 1863. The Paris metro opened in 1900, followed shortly by the New York subway. Eventually, highways, walkways, and high-speed trains were routed through tunnels, and electrical and fiber-optic cables coiled their way beneath many city streets. Today entire parallel worlds exist underground, powering and supporting the cities that rise above them.

No Drinking from the Tap

The most essential and the most easily overlooked miracle that sewer systems in part make possible is enjoying a glass of clean drinking water from a tap. Just a hundred and fifty years ago, in cities around the world, drinking water was very risky. You didn't know how contaminated the water was. Even Chicagoans, despite Chesbrough's brilliant plan, were in danger. There was literally a fatal flaw in their new sewer system.

Chesbrough's underground engineering successfully drained waste away from the city's streets, privies, and cellars. But almost all of his sewer pipes drained into the Chicago River, which emptied directly into Lake Michigan, the primary source of the city's drinking water. In the early 1870s, Chicago's water supply was so appalling that when you turned the taps on, your sink or tub would regularly fill with dead fish. They had been poisoned by the human filth in the river or lake and then sucked up into the city's water pipes. The locals called the disgusting watery mix "chowder."

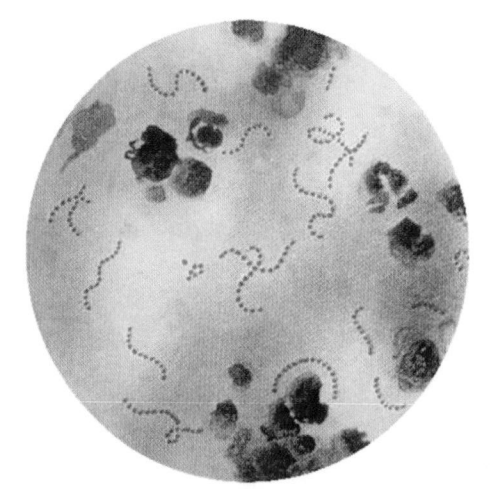

A Streptococcus pyogenes *bacteria (magnified nine hundred times) that causes puerperal or "childbed" fever. Ignaz Semmelweis, like most everyone else at the time, would not have known about these germs.*

The situation in Chicago was replicated around the world: sewer systems removed waste, but more often than not, sewer pipes simply poured it into the drinking water supply, either directly, or indirectly during heavy rainstorms. Most people were blissfully unaware that allowing waste to mix with drinking water was so deadly. No one could see the hidden killers that lurked in the dirty water. To create clean, healthy living conditions, we needed to understand what was happening on the scale of microorganisms. We needed to know how germs—whether they were viruses or bacteria—invaded our bodies and impacted our health and then discover a way to keep those germs from harming us.

Wash Your Hands

In 1847, a Hungarian physician, Ignaz Semmelweis, tried to connect the dots between behavior and disease. Semmelweis worked in the Vienna General Hospital in Austria, which had two maternity wards: one for well-to-do pregnant women who were attended by physicians and medical students, the other for working-class women who were cared for by midwives. Semmelweis noted that far fewer of the poorer women died from childbed fever after giving birth. He studied the practices in both maternity wards and made a

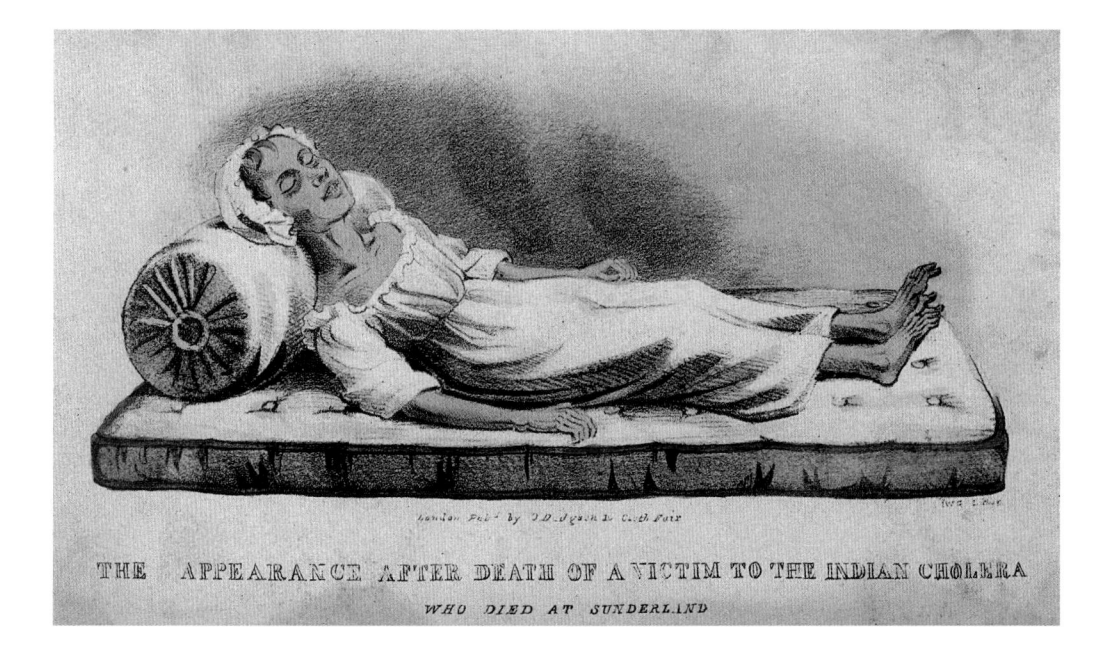

disconcerting discovery: the elite physicians and medical students switched back and forth from delivering babies in the wealthy ward and performing autopsies on cadavers in the morgue. Washing in between was minimal or nonexistent.

Semmelweis hypothesized that some kind of infectious particle was being transmitted by hand from the corpses to the new mothers. He ordered hospital staff to clean their hands and instruments with chlorinated lime to stop the cycle of infection. Semmelweis's physician peers were unenthusiastic; they didn't like being blamed for carrying diseases (in fact, they didn't like Semmelweis, who could be prickly). The medical establishment mocked, criticized, and ignored the idea of antiseptic cleansing, a fatal mistake for many of their patients as well as for Semmelweis, who was drummed out of the hospital and eventually died in a psychiatric institution.

This print from 1832 shows the effects of cholera. The 1800s saw five major cholera pandemics, and millions of people all over the world died horrible deaths from the disease, which still kills almost one hundred thousand people every year.

Lack of understanding about the relationship between germs and disease meant illnesses like cholera raged, especially in crowded urban centers with poor sanitation. Medical leaders of the time were still convinced that this illness spread through the stink in the air. But a cholera outbreak in London in 1854 that killed 616 people caused one man to rethink the miasma theory.

Beer versus Bacteria

John Snow was an English physician who had spent time in mines treating men stricken with cholera. He breathed the same putrid air they did, but didn't get sick. When he moved to London, he learned more about the way gases are distributed in the atmosphere. Snow began to question the miasma theory and formulated a new idea: cholera didn't come from the air; it was in the water.

Snow's 1854 cholera map of London.

The 1854 cholera outbreak in London began in the Soho area, a neighborhood with slaughterhouses, no sewers, and a dense population crammed into buildings with filthy cesspools underneath them. Snow knocked on doors and recorded the deaths at each address, mapping the locations and then assembling the data he gathered. A pattern emerged: the deaths were concentrated around a water pump on Broad Street that many people in the neighborhood used. But one group of Soho locals escaped the outbreak: the beer-drinking

John Snow, date unknown.

workers at the local brewery. Unbeknownst to them, the brewing process killed the disease. The brewery workers who drank beer weren't ingesting the organisms in contaminated water that Snow called "animalcules." This was the evidence he needed to support his waterborne theory of how diseases like cholera spread.

Snow never managed to see the bacteria that cause cholera directly; the technology of microscopy at the time made it almost impossible to see organisms that small. But he was right on the source of his animalcules: they weren't smells in the air; they were in the water. And if you were unlucky enough to drink that cholera-infected water, there would soon be upward of a trillion of these deadly creatures living in your gut.

Maps, Microscopes, and a Bath

Snow's work helped inaugurate the new science of epidemiology, using maps and surveys instead of lab-based experiments to uncover the patterns and causes of disease. Then a new innovation, in glass, made his theories even clearer.

In the early 1870s, the German lens crafters Zeiss Optical Works began producing new microscopes. For the first time, these devices were constructed around mathematical formulas that described the behavior of light. The new lenses enabled the microbiological work of German physician Robert Koch, one of the first scientists to identify the cholera bacterium. Along with his great rival, the French biologist-chemist Louis Pasteur, Koch and his microscopes helped develop and spread the germ theory of disease.

From a technological standpoint, the great nineteenth-century breakthrough in public health—the knowledge that invisible germs can kill—was a kind of team effort between maps and microscopes. Slowly, attitudes began to shift, most notably in England and America. Reformers built public bathhouses in urban slums where many people did not have access to indoor plumbing. A minor genre of self-help books and pamphlets emerged, with detailed instructions teaching people how to take a bath. It may seem strange to think that people needed to read a book to learn how and why to bathe, but until the 1800s, Europeans and Americans thought that submerging the body in water was unhealthy and that clogged pores protected you from disease. In fact, baths were so repugnant to people back then that even the wealthiest members of society avoided them at all costs. Queen Elizabeth I bothered to take a bath only once a month, and she was a veritable clean freak compared to her peers. The French king Louis XIII was not bathed once until he was seven years old.

Personal hygiene and the availability of clean water were directly linked, but authorities needed quantifiable information in order to make important decisions about large-scale public sanitation. They got it, thanks once again in part to the pioneering efforts of Robert Koch. The new lenses and tools for microbiological research meant Koch could now see and *measure* bacteria. He experimented with mixing contaminated water with transparent gelatin

The cover of Catherine Beecher and Harriet Beecher Stowe's influential book about health and hygiene, 1869. Stowe is more widely known as the author of Uncle Tom's Cabin.

and viewing the growing bacterial colonies on a glass plate. He developed sophisticated tools to measure the density of bacteria in water. In the 1880s, Koch established a unit of measure that could be applied to any amount of water: if there were fewer than one hundred bacterial colonies per milliliter, it was safe to drink.

The ability to measure bacterial content was a real breakthrough in meeting the challenges of public health. Before the adoption of these measurement standards, any improvements to a water system were tested the old-fashioned way: you built a new sewer or reservoir or pipe, and then you sat around and waited to see if fewer people died. The ability to take a sample of water and determine scientifically whether it was free of contamination accelerated experimentation and new designs in sanitation. As public infrastructure improved, people were much more likely to have running water in their homes, and that water was cleaner than it had been a few decades earlier. Most important, the germ theory of disease had gone from fringe idea to scientific consensus.

Shh! It's a Secret!

Microscopes and measurement quickly opened a new front in the war on germs: instead of fighting bacteria indirectly by routing waste away from drinking water, new chemicals could be used to attack the germs directly. One of the key soldiers on this second front was a New Jersey doctor named John Leal.

Leal's interest in public health, especially contaminated water supplies, was born of a personal tragedy. His father had suffered a slow and painful death from drinking bacteria-infested water during the Civil War.

In the late 1890s, Leal experimented with many techniques for killing bacteria, but one poison in particular piqued his interest: calcium hypochlorite, known at the time as "chloride of lime" (Semmelweis's cleanser of choice). Chloride of lime, or chlorine, was by then in wide circulation as a public health remedy. Houses and neighborhoods that had suffered an outbreak of typhoid or cholera were routinely disinfected with the chemical, an intervention that did nothing to combat waterborne disease. Its sharp, acrid odor was indelibly associated with epidemics in the minds of many Americans and Europeans. Plus, chloride of lime is potentially lethal. So the idea of putting chlorine in water had not yet taken hold, and most doctors and public health authorities rejected the approach.

Armed with tools that enabled him to both see and measure the microorganisms behind waterborne diseases, Leal became convinced that adding chlorine, at the right dosage, would rid water of dangerous bacteria more effectively than any other means, with no threat to the humans drinking it. But how to prove this on a mass scale? When Leal landed a job with the Jersey City Water Supply Company, he found the answer . . . in secret.

Leal now had oversight of seven billion gallons of water in the Passaic River watershed, which provided the drinking water for the two hundred thousand inhabitants of Jersey City, New Jersey. They became unwitting participants in one of the most bizarre and daring interventions in the history of public health. In almost complete secrecy, with no permission from government authorities and no notice to the general public, John Leal chlorinated their water.

The Jersey City Water Supply Company was fighting the city in several lawsuits about the quality of water they provided. The company was ordered to build new sewer lines, which Leal knew wouldn't solve the problem. So he went rogue.

Engineer George Warren Fuller helped Leal build and install a "chloride of lime feed facility" at the reservoir outside Jersey City. The plant went online on September 26, 1908. It was the first mass chlorination of a city's water supply in history and a staggering risk, given the popular opposition to chemical filtering.

Word got out about Leal's bold experiment three days later, and he was branded by some as a madman or terrorist. Ordered to appear in court to defend his actions, he testified that he would drink the chlorinated water himself and not hesitate to pour some for his family, and that it was the safest water in the world. The court case was settled. Within a few years, the data supported both Leal's daring move and his theory: Jersey City saw dramatic decreases in waterborne diseases.

Leal chose not to patent his chlorination technique so that it could be freely adopted by any water company and municipality. Researchers found that between 1900 and 1930, when chlorination and other water filtration techniques were first rolled out across America, clean drinking water led to a 43 percent reduction in mortality in the average American city; infant mortality dropped by an incredible 74 percent. Chlorination became standard practice across the United States and eventually around the world.

Chlorination wasn't just about saving lives, though. It was also about having fun. After World War I (1914–18), ten thousand chlorinated public baths and pools opened across America; learning how to swim became a rite of passage. After World War II (1939–45), America's swimming craze moved into the private sphere. By the 1960s, millions of U.S. homes had chlorinated swimming pools.

Clean, Cleaner, Cleanest

Through the nineteenth century, the advance of clean technologies had mainly brought about big public health projects and systems. But the story of clean in the twentieth century is a much more intimate and personal affair.

Just a few years after Leal's bold experiment, five San Francisco entre-

Bathing suit contest, 1921. Chlorine made public swimming pools possible. That triggered a seemingly unrelated long-zoom change: women's fashions. At the beginning of the twentieth century, the average woman's bathing suit required ten yards of fabric; by the end of the 1930s, one yard was enough.

preneurs invested a hundred dollars each to launch a chlorine-based product. It seems with hindsight to have been a good idea, but their bleach business was aimed at big industry, and sales didn't develop as quickly as they hoped. But the wife of one of the investors, Annie Murray, a shop owner in Oakland, California, thought that chlorine bleach could be a revolutionary product for people's homes as well as factories.

At Murray's insistence, the company created a weaker version of the chemical and packaged it in smaller bottles. Murray was so convinced of the product's promise that she gave out free samples to all customers at her shop.

Within months, bottles were selling like crazy. Murray didn't realize it at that time, but she was helping to invent an entirely new industry.

Annie Murray had created America's first commercial bleach for the home, and the first in a wave of cleaning brands that would become ubiquitous in the new century: Clorox.

Hygiene products for the home were among the first to be promoted in full-page advertisements in magazines and newspapers. The 1920s ushered in an era where Americans were bombarded by commercial messages convincing them that they faced certain humiliation if they didn't do something about the germs on their bodies or in their homes. Soap, bleach, mouthwash, and deodorant would save you: buy now!

Radio, and later television, stations began experimenting with storytelling; in a brilliant marketing move, the ads of personal-hygiene companies supported these broadcasts. The modern-day "soap opera" was born.

Today the cleaning business is worth an estimated eighty billion dollars. Walk into a big-box supermarket or drugstore and you will find hundreds, if not thousands, of products devoted to ridding our households of dangerous germs or scouring ourselves from teeth to feet.

The conflict between humans and bacteria that played out over the past two centuries and our growing understanding of the microbial routes of disease have had far-reaching consequences. As of 1800, no society had successfully built and sustained a city of more than two million people; clean drinking water and reliable waste removal changed that. Now, the populations of London and New York City each top eight million, and life expectancies are longer, disease rates lower.

However, 40 percent of the world's population today—close to three billion people—still lacks access to clean drinking water and basic sanitation systems. Many of those people live in the poorer sections of megacities like Delhi, India (population twenty-four million), that lack infrastructure as

A 1936 ad for Clorox.

well as basic services. So the question now is, how do we broaden the clean revolution? How can we use new ideas and technology to inspire new solutions, the same way that germ theory and the microscope triggered the hunch about chemically treating water? And can we even bypass the costly, labor-intensive work of building giant infrastructure or engineering projects?

In 2011, the Bill and Melinda Gates Foundation announced a multiyear competition to help generate a shift in the way we think about basic sanitation services. The Reinvent the Toilet Challenge solicited designs for toilets that do not require a sewer connection or electricity and cost less than five cents per user per day. The first-prize-winning entry was a toilet system from the California Institute of Technology. In Caltech's self-contained design, a solar panel powers an electrochemical reactor that treats human waste, producing clean water for flushing or irrigation and hydrogen that can be stored in fuel cells. Inexpensive computer chips monitor the system. The toilet is being tested in two sites in India and one in China.

Clearly, high tech can be used to produce something clean and easy to use in a low-tech way. Interestingly enough, computer technology is itself a by-product of the clean revolution. Without clean water, we wouldn't have smartphones and laptops.

Computer chips are fantastically intricate creations. Their microscopic detail is almost impossible to comprehend. To measure them, we need to zoom down to the scale of micrometers, or microns: one-millionth of a meter. Manufacturing at this scale involves extraordinary robotics and laser tools. But microchip factories also require another kind of technology: they need to be preposterously clean. In fact, they are some of the cleanest places on the planet. A speck of household dust landing on one of the delicate silicon wafers these factories produce would be like Mount Everest falling on a street corner in Manhattan.

Microchip plants also require ultrapure water, and plenty of it, to use as

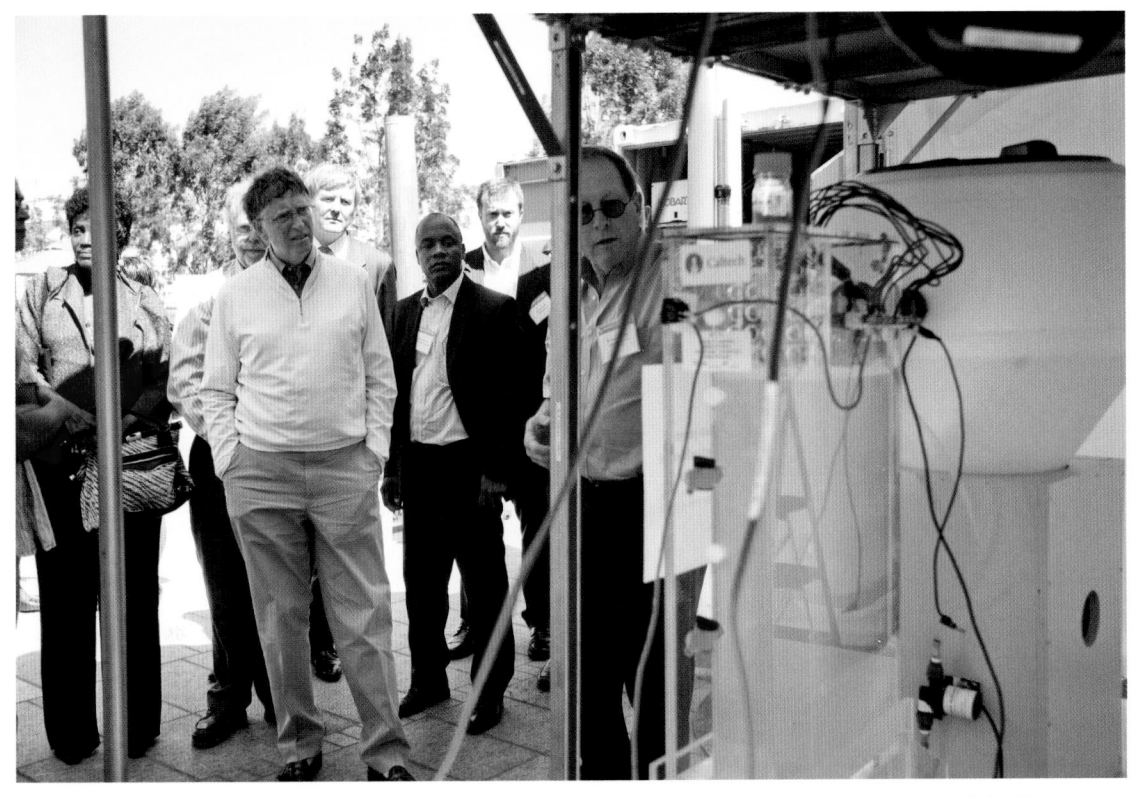

Bill Gates (left) checks out Caltech's model. Other research groups are also designing "toilets of the future" and in 2013 the United Nations General Assembly declared November 19 as World Toilet Day to raise awareness about the problems of global sanitation.

a solvent for the microchips. To avoid impurities, any bacterial contaminants *and* all minerals, salts, and random ions are filtered from the water. This process creates pure H_2O, which is undrinkable; chug a glass of the stuff and it will start leeching minerals out of your body.

This is the full circle of clean: some of the most brilliant ideas in science and engineering in the nineteenth century helped us purify water that was too dirty to drink. And now, 150 or so years later, we produce water that's too *clean* to drink in order to manufacture the tools and tech we need for our digital age—including the "toilets of the future"!

Time

What time is it?

If you had asked this question 150 years ago, you would have received at least twenty-three different answers in the state of Indiana, twenty-seven in Michigan, and thirty-eight in Wisconsin. It all depended on where you were and how you were keeping time.

This circa-1856 photo shows Egypt's Karnak temple complex, which dates from the 1500s–1400s BC. Egyptians marked time using the shadows of the two obelisks. Their system influenced how we came to divide time into shorter units of seconds, minutes, and hours.

For almost the entire span of human history, time had been calculated by tracking the heavenly rhythms of solar bodies. Days were defined by the cycle of sunrise and sunset, months by the cycles of the moon, years by the slow but predictable rhythms of the seasons. For most of that history, we misunderstood what was causing those patterns, assuming that the sun revolved around the earth and not the reverse.

Back and Forth and Back and Forth . . .

Over time, people built tools to measure the passing of the day and eventually of the hours, minutes, and seconds. One of those people was a student in sixteenth-century Italy whose daydreaming led to an essential discovery.

Pisa, Italy, is famous for its leaning cathedral tower and its learned native son Galileo Galilei. Suspended from the ceiling of Pisa's thousand-year-old cathedral is a collection of altar lamps. Legend has it that in 1583, nineteen-year-old Galileo was daydreaming in the pews during prayers and noticed one of the altar lamps swaying back and forth. He became almost hypnotized by the lamp's regular motion. No matter how large the arc, the lamp appeared to take the same amount of time to swing back and forth. As the arc decreased in length, the speed of the lamp decreased as well. To confirm his observations, Galileo measured the lamp's swing against the only reliable timekeeper he could find: his own pulse.

Galileo spent the next twenty years demonstrating his genius in physics, mathematics, and astronomy. He experimented with telescopes, and more or less invented modern science. But still, the altar lamp swung back and forth in his mind. He decided to build a pendulum that would re-create what he had observed in the cathedral in Pisa. Galileo discovered that the time it takes a pendulum to swing is dependent upon the length of string on which an object hangs, not the size of the arc or the object's mass. "The marvellous

The overhead lamps at Pisa Cathedral that inspired Galileo.

property of the pendulum," he wrote to a fellow scientist, "is that it makes all its vibrations, large or small, in equal times."

In equal times. In Galileo's age, machines that kept a reliable beat didn't exist. Most Italian towns had large, unwieldy mechanical clocks that had to be corrected by sundial readings constantly or they lost or gained as much as twenty minutes each day. It was a challenge for state-of-the-art timekeeping technology in the Renaissance to stay accurate on the scale of *days*—but no one really noticed.

Wait, Where Are We?

In the middle of the sixteenth century, most people had no need for split-second accuracy. If you knew roughly what hour of the day it was, you could get by just fine . . . as long as you stayed on land. But this was the first great age of global navigation. European explorers, inspired by Columbus, were sailing to the Far East and the Americas in search of fortune and fame. And if they were going to survive, they definitely needed time on their side. Navigators could calculate their latitude—how far north or south they were—by charting the sun and other stars. But to chart longitude—how far east or west they were—they needed clocks.

Before modern navigation technology, the only way to figure out a ship's longitude was to have two clocks on board. One clock was set to the exact time of your origin point (assuming you knew the longitude of that location). The other clock recorded the current time at your location at sea. The difference between the two times gave you your longitudinal position: every four minutes of difference translated to one degree of longitude, or sixty-eight miles at the equator.

In clear weather, you could easily reset the ship's location clock through accurate readings of the sun's position. The problem was the port of origin clock. With timekeeping technology losing or gaining up to twenty minutes each day, the port clock was practically useless by day two of the sea journey.

All across Europe, lucrative bounties were offered for a solution to the problem of how to determine longitude at sea. Galileo worked on the problem, but his method based on astronomical observations was too complicated to use and not accurate enough. His thinking swung back to his pendulum, in the hope

Galileo was seventy-seven years old and blind when he returned to the idea of a pendulum clock, so his son drew this design in 1641. The pendulum's swing creates equal beats, which control the hands of the clock. Galileo never got a chance to test his revolutionary idea at sea.

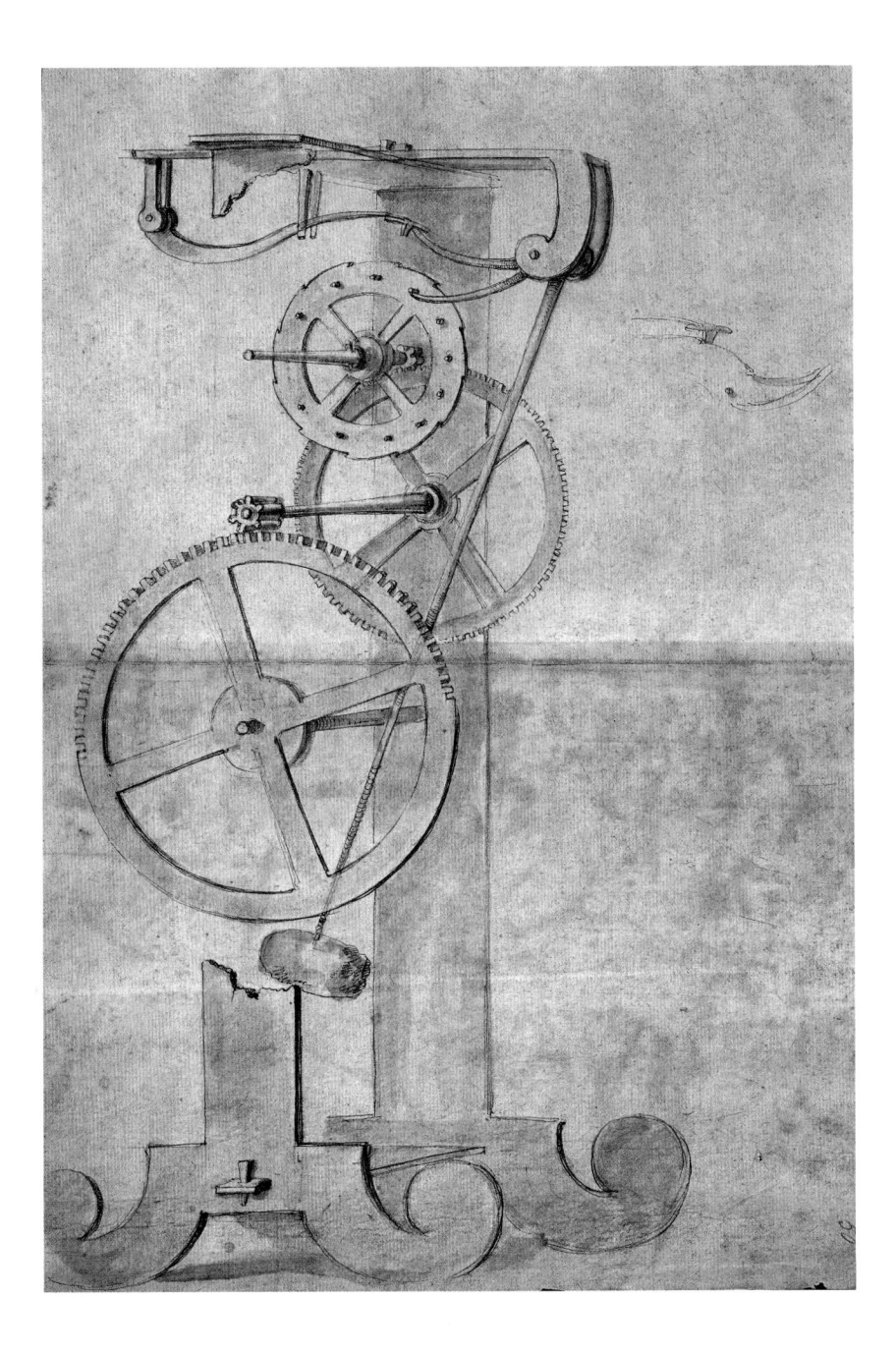

that his slow hunch about its "marvelous property" might pay off. With the help of his son, Galileo drew up plans for the first pendulum clock in 1641. Though the longitude problem wouldn't actually be solved until 1761, when Englishman John Harrison invented his precision marine chronometer, Galileo's pendulum clock was a significant advancement in timekeeping technology.

The Pendulum Swings

By the 1700s, the pendulum clock had become a regular sight in workplaces, town squares, and the homes of the rich throughout Europe, particularly in England. It only lost or gained a minute or so each week, which made the pendulum clock about a hundred times more accurate than earlier time-pieces. From the long-zoom angle, the pendulum clock didn't just *tell* time; it changed how we *experienced* time.

Think about the early days of the Industrial Revolution in England in the middle of the eighteenth century. Does noise come to mind? As in, thundering steam engines, clacking steam-powered looms, and other clanging, clattering machinery? Beneath all this factory cacophony, there was another important sound: the ticking of pendulum clocks, quietly keeping time.

Accurate clocks were essential to the industrialization of the world. They determined longitude at sea and greatly reduced the risks of global shipping networks. Safer shipping meant a steady supply of raw materials for industrialists, who could then ship their goods to markets overseas. In the late 1600s and early 1700s, the most reliable watches in the world were manufactured in England, which created a pool of expertise with fine-tool manufacture. That pool would prove to be incredibly handy when the demands of

The German-made Riefler pendulum clock (shown here) was accurate to a hundredth of a second a day. The National Institute of Standards and Technology, a U.S. government agency, used it as the standard for time measurement from 1904 to 1929.

industrial innovation arrived, just as the glassmaking expertise in producing spectacles opened the door for telescopes and microscopes. The watchmakers were the advance guard of what would become industrial engineering.

The Time of Your Life

More than anything else, clocks were needed to move people from organic time to organized time. When people primarily farmed or lived in rural communities, time was usually described in relation to how long it took to complete a task. You'd talk about meeting up in the time it took to milk a cow, not specific minutes. Instead of being paid by the hour, craftsmen were paid by the piece produced. People's schedules were varied and unregulated. Clearly, that wasn't going to work once the factory whistles started blowing.

When people left the farm fields for the factory floor, our experience of time changed forever. You had to arrive on time and clock in for fourteen-hour shifts. Your workday didn't follow the sun; you woke up and left home in the dark and kept going until it was dark out again. If your body clock needed help with that, you had a cup of coffee or tea. A whole new business in stimulants helped keep you awake and working.

For the first generations living through the Industrial Revolution, "time discipline" was a shock to the system. The pendulum clock took the informal flow of life and regulated it with a mathematical grid. Once again, an increase in our ability to measure things turned out to be as important as our ability to make them.

But not everyone owned a clock; factory owners hired "wakers" to rouse laborers so they'd get to work on time. Pocket watches remained luxury items until the middle of the nineteenth century. At the time, producing a watch involved more than a hundred distinct jobs, from turning a piece of steel on a thread to make individual flea-sized screws to inscribing watchcases.

Workers wait to punch the time clock at a Philadelphia factory, circa 1942. The industrial age introduced new ideas about time, including time clocks and hourly wages.

That changed when a Massachusetts cobbler's son, Aaron Dennison, borrowed from the new process of manufacturing weapons, which used standardized, interchangeable parts, and applied the same techniques to watchmaking. Dennison's "Wm. Ellery" watch—named after William Ellery, one of the signers of the Declaration of Independence—turned a luxury item into a must-have gadget for a mainstream audience. In 1850, the average pocket watch cost $40; by 1878, a Dennison watch cost just $3.50.

What Time Do You Have?

Thanks to Dennison, there were now watches in plenty of pockets—but they were all running at different times. In the United States, each town and village synced its clocks to the sun's position in the sky. If you moved west or east by even a few miles, the shifting relationship to the sun would produce a

different time on a sundial. You could be standing in one city at 6:00 p.m., but just three towns over, the correct time would be 6:05. Clock time had been democratized, but it had not yet been standardized.

The strangest thing about this irregularity is the fact that no one noticed it. You couldn't talk directly to someone three towns over, and it took an hour or two to get there by unreliable roads at low speeds. So a few minutes of fuzziness in the respective clocks of each town didn't even register. But once people and information began to travel faster, it became a massive problem. Telegraphs and railroads exposed the hidden blurriness of nonstandardized clock time, just as centuries earlier the invention of the printing press had exposed the need for spectacles.

Portrait of an unidentified solider with a pocket watch, circa 1860. During the Civil War, more than 160,000 of Aaron Dennison's pocket watches were sold. Even Abraham Lincoln carried a "Wm. Ellery."

Trains moving east or west play tricks with our measurement of time. Traveling west makes time seem to progress more slowly because you are keeping up with the sun as it appears to move across the sky. Traveling east has the opposite effect. For every hour you traveled on a train, you needed to adjust your watch by a few minutes, depending on how fast the train was going. You also had to factor in that each railroad ran on its own clock. Going on a journey in the nineteenth century took some formidable number crunching.

The British had dealt with a similar problem in the 1840s by standardizing the entire country on Greenwich Mean Time. (Greenwich, England, is located on the prime meridian, where the longitude is zero degrees.) Railroad clocks were synced by telegraph. But the United States was too sprawling to run off of one clock, particularly after the transcontinental railroad opened in 1869. With eight thousand towns across the country, each on its own clock, and more than a hundred thousand miles of railroad track connecting them, the need for some kind of standardized system was overwhelming.

In the early 1880s, a railroad engineer named William F. Allen took on the cause. Allen was the editor of a guide to train timetables, so he knew firsthand how exasperating the existing time system was. At a railroad convention in St. Louis in 1883, Allen presented a map that shifted fifty distinct railroad times to four time zones: Eastern, Central, Mountain, and Pacific. Allen designed the map so that the divisions between time zones zigzagged slightly to correspond to the points where the major railroad lines connected, instead of running straight down longitudinal lines.

The railroad bosses were convinced, and gave Allen nine months to make his idea a reality. He launched an energetic and ultimately successful campaign of letter writing and arm-twisting to bring politicians and other officials on board. And so November 18, 1883, became "the day of two noons" in

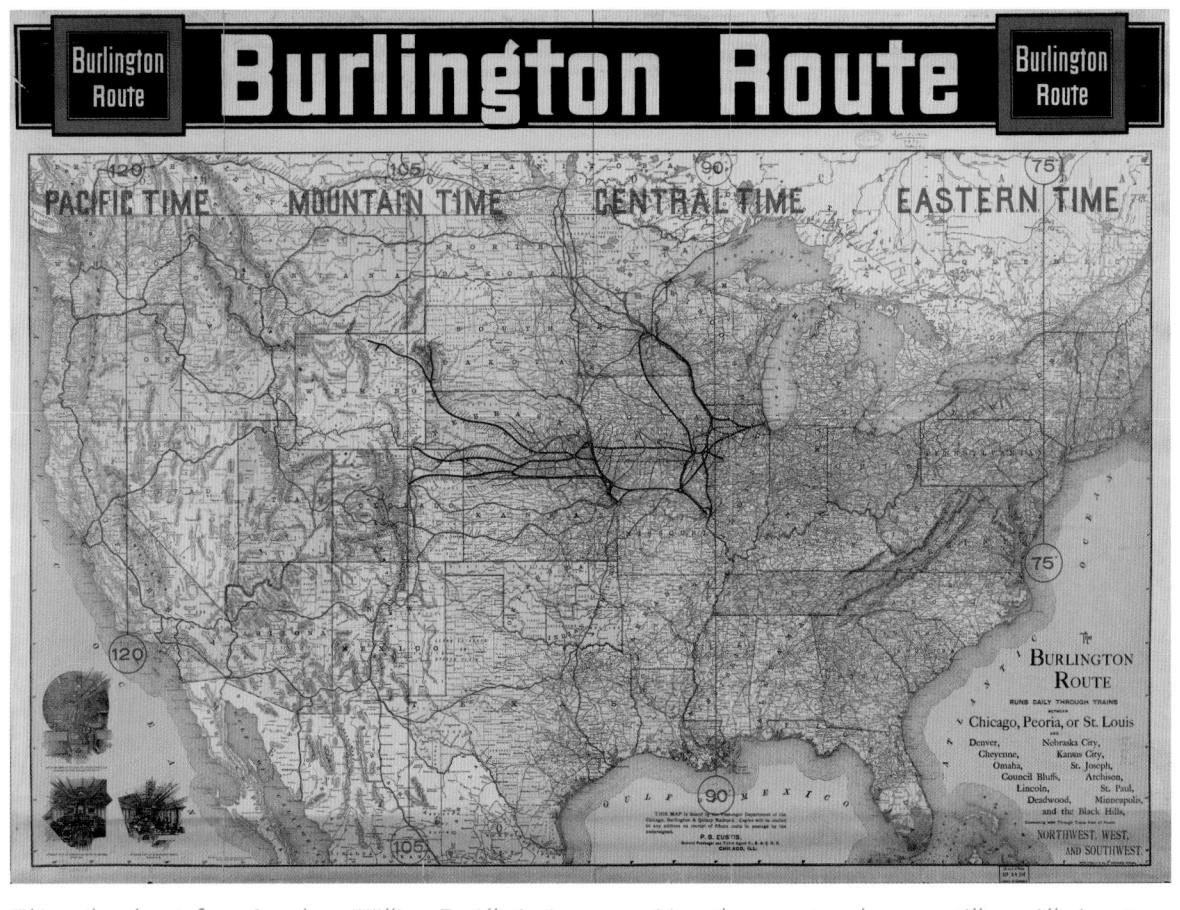

This railroad map from 1892 shows William F. Allen's time zones. More than a century later, we still use Allen's system.

the United States. Allen's Eastern Standard Time ran exactly four minutes behind what was then local time in New York City. So on that unusual November day, the church bells in Manhattan pealed at the old New York noon, and then four minutes later rang out a second noon, the very first 12:00 p.m., Eastern Standard Time (EST). This second noon was broadcast across the country via telegraph, allowing railroad lines and town squares all the way to the Pacific to synchronize their clocks according to their respective time zones.

The following year, Greenwich Mean Time became the international clock, and the whole globe was divided into time zones. World trade, travel, and communications greatly improved once people could sync their clocks. The next revolution in time would involve synchronizing devices even more accurately.

Good Vibrations

In the 1880s, French scientists Pierre and Jacques Curie detected something interesting about crystals such as quartz: if you applied enough pressure, you could make them vibrate, especially if you zapped them with an alternating current. Even more important, the quartz crystal shook at a steady 32,768 times per second—a regular movement, just like Galileo's swinging lamp. And just as Galileo's observations led to the pendulum, the Curie brothers' findings on what is known as piezoelectricity gave rise to a new kind of clock.

In 1928, W. A. Marrison and J. W. Horton of Bell Labs built the first quartz clock; it used the crystal's regular vibrations to keep time. This quartz clock lost or gained only a thousandth of a second per day. Compared to a pendulum clock, it was far less vulnerable to changes in temperature or humidity, not to mention movement.

Quartz clocks became the main timekeepers for science and industry; starting in the 1930s, standard U.S. time was kept by quartz clocks. By the 1970s, mass-market, quartz-based wristwatches were widely available. Today, just about every consumer appliance that has a clock—computers, cell phones, tablets, microwaves, alarm clocks, wristwatches, automobile clocks—runs on quartz piezoelectricity.

Quartz time also made something possible that wouldn't at first seem to have anything to do with time: computation. Computer microprocessors

The first quartz clock, circa 1928. The cut of a quartz crystal determines how many times per second it can vibrate. Marrison and Horton's invention used high-frequency quartz vibrations: 50,000 per second.

execute billions of calculations per second, while shuffling information in and out of other microchips on their circuit boards. These operations are all coordinated by a master clock, now almost without exception made of quartz. The fact that there are computers for doing homework or watching videos is not just due to innovators like Steve Jobs and Bill Gates; our computers are also dependent on centuries of innovations from clockmakers.

Out of Astronomical Time

Once we started measuring days with quartz clocks, we discovered that the length of the day was not as reliable as we had thought. Days shortened or lengthened in semi-chaotic ways, thanks to the drag of the tides on the surface of the planet, wind blowing over mountain ranges, or the inner motion of the earth's molten core. If we really wanted to keep exact time, we couldn't rely on the earth's rotation. We needed a better timepiece.

The discovery of the atom in the early days of the twentieth century—spearheaded by scientists such as Niels Bohr and Werner Heisenberg—led to some spectacular and deadly innovations in energy and weaponry, including nuclear power plants and hydrogen bombs. Atomic research also revealed a less celebrated but equally significant discovery. Bohr noticed that the electrons orbiting within a cesium atom moved with an astonishing regularity. These electrons tapped out a rhythm that was far more reliable than the earth's rotation, and could be used to measure equal intervals of time.

The first atomic clocks were built in the mid-1950s, and they immediately set a new standard: we could now measure a nanosecond, a *billionth* of a second. Atomic clocks are a thousand times more accurate than quartz clocks. They can measure time with a drift of only a single second every *five billion years*!

On October 13, 1967, the International Conference of Weights and Measures declared that the master time for the planet would be measured in atomic seconds. A day was no longer the time it took the earth to complete one rotation. A day became 86,400 atomic seconds, ticked off on 270 synchronized atomic clocks around the world. These clocks use quartz mechanisms and are reset every year so that the atomic and solar rhythms don't get too far out of sync.

Atomic time has radically transformed everyday life. Global air travel, telephone networks, and financial markets all rely on nanosecond accuracy. Whenever you glance down at your smartphone to check your location, you are consulting a network of twenty-four atomic clocks housed in satellites in low Earth orbit. Those satellites are sending out signals, again and again: *The time is 11:48:25.084738. . . . The time is 11:48:25.084739. . . .* Since the satellites are in predictable positions, a truly "smart" phone can calculate its exact position by triangulating among three different time stamps. This Global Positioning System (GPS) is the modern high-tech version of what the naval navigators of the eighteenth century did: compare clocks.

Long, *Very Long*, Time

The story of time's measurement seems to be all about acceleration, dividing up the day into smaller and smaller increments so that we can move things faster. But time measurement in the atomic age has also moved in the exact opposite direction: measuring in eons, not microseconds.

In the 1890s, Polish-French scientist Marie Curie proposed that radiation was not some kind of chemical reaction between molecules, but something intrinsic to the atom—a discovery so critical to the development of physics that she later became the first woman to win a Nobel Prize. Her research quickly drew the attention of her husband, Pierre Curie, who abandoned his own studies of crystals to focus on radiation. Together they discovered that radioactive elements decayed at constant rates. The half-life of carbon-14, for instance, is 5,730 years. Leave some carbon-14 lying around for five thousand years or so, and you'll find that half of it is gone.

Once again, science had uncovered a new source of "equal time," only this clock wasn't counting out the microseconds of quartz oscillations or the nanoseconds of cesium electrons. Radiocarbon decay was ticking on the scale

A 1904 French magazine featured Marie and Pierre Curie working in their lab on its cover.

of centuries or millennia, and this decay rate could also be used as a "clock."

Most clocks focus on measuring the present: what time is it right now? But radiocarbon clocks are all about the past. Different elements decay at wildly different rates, which means that they are like clocks ticking at different time scales. Carbon-14 "ticks" every 5,730 years, but potassium-40 "ticks" every 1.3 billion years. Carbon dating, as the process is known, is an ideal clock for the "deep time" of human history, while potassium-40 measures geologic time, the history of the planet itself. When *Homo sapiens* first crossed into the Americas more than ten thousand years ago, there were no historians capable of writing down a narrative account of their journey. Yet the story was captured by the carbon in the bones of these early people and the charcoal deposits they left behind at campsites. We have immense knowledge about the prehistoric migrations of humans across the globe in large part thanks to carbon dating. Without it, the deep time of human migrations or geologic change would be like a history book where all the pages have been randomly shuffled: teeming with facts but lacking chronology and causation.

This is the peculiar paradox of time in the atomic age: we live in ever-shorter increments, guided by clocks that tick invisibly with immaculate precision; we have short attention spans and have surrendered our natural rhythms to the abstract grid of clock time. And yet simultaneously, we have the capacity to imagine and record histories that are thousands or millions of years old, to trace chains of cause and effect that span dozens of generations. Our time horizons have expanded in both directions, from the microsecond to the millennium. The long-zoom knowledge that we've gained can help us tackle some of the twenty-first century's most important problems.

The first prototype of the "clock of the Long Now." Inside a mountain in western Texas an enormous clock is being built. It will be about two hundred feet tall and tick for ten thousand years, roughly the length of human civilization to date. The Long Now Foundation, the visionaries behind this marvel, hope it will expand our sense of time, especially about present actions and future consequences.

Light

·········▶

Imagine some alien civilization observing Earth, looking for signs of intelligent life. For millions of years, there would be almost nothing to report. But starting about a century ago, a momentous change would suddenly be visible: at night, the planet's surface would glow with the streetlights of cities, first in the United States and Europe, then spreading steadily across the Earth. Viewed from space, the emergence of artificial light would be the single most dramatic change in our planet's appearance in the last sixty-five million years.

This view from the International Space Station shows the eastern United States at night. Today's night sky shines six thousand times brighter than it did just one hundred fifty years ago.

Artificial light has transformed the way we live, work, and even sleep. It has inspired innovations that impact every aspect of our lives, from manufacturing and architecture to home goods and entertainment. It's helped create global networks of communication and is at the core of developments in energy production. But as bright ideas go, artificial light was a slow burn.

Candlelight Wasn't Always Romantic

More than a hundred thousand years ago, humans mastered the controlled fire, which first provided artificial light. While some ancient cultures used oil lamps, for thousands and thousands of years, the humble candle warded off the dark—in its own short-lived, smelly way. Candles made from beeswax were highly prized but too expensive for anyone except the wealthy. Most people made do with tallow candles, which burned animal fat to produce flickers of light, along with thick smoke and a strong, foul odor. In many ordinary households, tallow candles were homemade. You heated, stirred, and skimmed vats of cow or sheep fat and repeatedly dipped woven wicks in the stinky concoction or used a combination of wicks and molds. The tallow, or fat, built up around the wick and then hardened. Candle making could go on for days, since the average colonial home in America used about four hundred candles a year.

Just imagine you're a New England farmer in deep winter in the 1700s. The sun sets at five o'clock, and you're facing many hours of darkness. It's pitch-black: there are no streetlights, flashlights, lightbulbs, fluorescents—even kerosene lamps haven't been invented yet. There's just the flickering glow of a fireplace and the smoky burn of the tallow candle.

Might as well go to bed early.

Scientists now believe that before there was widespread artificial lighting, people's sleep patterns were radically different from ours. When darkness fell, they drifted into "first sleep" and woke after four hours or so to snack, relieve themselves, have sex, or chat by the fire. They then headed back to bed for four hours of "second sleep." Flickering candlelight wasn't strong enough to transform our sleep pattern from two stages to one. A change that momentous required the steady, bright glow of nineteenth-century lighting. And a source for that kind of light was discovered in a most unpleasant place: the skull of a fifty-ton marine mammal.

Thar She Blows—and Burns!

Legend has it that sometime around 1712, a powerful storm off the coast of the island of Nantucket in Massachusetts blew a ship captain by the last name of Hussey (his first name is lost to history) and his ship far out to sea. In the deep waters of the North Atlantic, he encountered a species of whale that had never been seen before, a giant leviathan of the deep: the sperm whale.

THE SPERMACETI WHALE

A hand-colored nineteenth-century engraving of a sperm whale hunt. The harpoonist (right) had to get very close to the huge animal. Even harpooned, the powerful whale might take off, dragging the small whaling rowboat along on a "Nantucket sleigh ride."

Most believe Hussey harpooned the beast, and when the huge mammal was dissected, whalers found a cavity inside the creature's massive head. It was filled with a white, oily substance that resembled semen, or sperm. This whale oil came to be called "spermaceti."

Scientists are still not entirely sure why sperm whales produce spermaceti. Some believe the whales use the fluid for buoyancy; others think it helps

GRAND BALL GIVEN BY THE WHALES IN HONOR OF THE DISCOVERY OF THE OIL WELLS IN PENNSYLVANIA.

In this cartoon from 1861, relieved whales celebrate the oil industry overtaking the spermaceti business, which happened very quickly once fossil fuel–based options like kerosene became readily available in the 1860s. Whale oil continued to be used to light lamps and lubricate steam engines and locomotives in the 1800s, and it was used in the gear boxes of American automatic cars until 1972. Sperm whales' current conservation status is endangered.

with the mammal's echolocation system. But either way, you're talking about a lot of spermaceti: a mature sperm whale holds as much as five hundred gallons inside its skull. And those New Englanders soon figured out what to do with it.

They made the happy discovery that candles made from spermaceti produce a much stronger, whiter light than tallow candles, and aren't as smoky. By the second half of the eighteenth century, spermaceti candles had become the most prized form of artificial light in America and Europe.

The candle business became so lucrative that a group of manufacturers formed an organization called the United Company of Spermaceti Chandlers. This "Spermaceti Trust," as it was known, kept competitors out of the business and forced whalers to hold their prices in check. In other words, the organization was a price-fixing monopoly, one of the first on record.

Whaling was a dangerous and repulsive business, especially when it came to the extraction of spermaceti. Sailors carved a hole in the top of the head of the captured, dead sperm whale. Often they lowered the smallest person on board, perhaps a young cabin boy, into the cavity above the whale's brain to start bailing out the spermaceti. Eventually more men crawled into the giant head, spending days inside the smelly, rotting carcass hauling buckets of the valuable oil. It is strange to think that if your great-great-grandfather wanted to read his book at night, some twelve-year-old had to crawl around in a whale's skull for an hour!

In just over a century, nearly 300,000 sperm whales were slaughtered (today's sperm whale population is believed to range somewhere between 200,000 to 1.5 million, though it's almost impossible to get an accurate estimate). The entire population might have been killed off had we not found a new source of fuel for artificial light. This time the oil was in the ground.

Say Cheese!

Fossil fuels like coal, oil, and gas are formed in the earth from the ancient remains of once-living organisms. The first commercial use of fossil fuels revolved around artificial light. New kerosene and gas lamps that were twenty times brighter than any candle were invented in the mid-nineteenth century. Their superior brightness helped ignite many changes, including an explosion in magazine and newspaper publishing in the second half of the nineteenth century. Now that people could see in the evening hours, they wanted more material to read. And one man in particular, Jacob Riis, wanted people to know about what was happening in the darker corners of life. Innovations in artificial light helped him grab the public's attention.

In the 1830s, photographers still depended upon natural sunlight in order to take pictures. This limited their ability to shoot in the dark, or inside, or any time or place with insufficient natural light. Various photographers experimented with combustible chemicals and materials to provide artificial lighting that would expand the possibilities of photography. In the 1880s, two German scientists, Adolf Miethe and Johannes Gaedicke, mixed fine magnesium powder with potassium chlorate. When their concoction exploded, it gave off white light that allowed high-shutter-speed photographs to be taken in low-light conditions. They called the mix *Blitzlicht*: "flash light."

In October 1887, a New York newspaper ran a four-line dispatch about *Blitzlicht*. Jacob Riis, a young Danish immigrant who was a police reporter and amateur photographer, spotted it. Riis had long had a slow hunch about the need to expose the squalor of nineteenth-century tenement life in order to inspire reform. He'd been exploring the depths of Manhattan slums for years, but his journalistic reports about the appalling conditions he saw there had

Jacob Riis, circa 1900.

The interior of a New York City tenement photographed by Jacob Riis, 1888. With over half a million people living in only fifteen thousand buildings, some neighborhoods of Lower Manhattan were the most densely populated places on the planet.

failed to change public opinion in any meaningful way. So had exhaustive surveys with titles like "The Report of the Council of Hygiene and Public Health." Riis suspected that the problem with tenement reform—and urban poverty initiatives in general—was ultimately a problem of imagination. Americans, especially most voting Americans, were too far removed from the squalor and suffering to really imagine it, much less demand change. They just didn't get the picture—and that's what Jacob Riis wanted to give them.

Urban tenements were notorious for their lack of fresh air and light, even indirect sunlight. And this was the great stumbling block for Riis's photography—and maybe where *Blitzlicht* could shine light in the darkness.

Riis assembled a team of amateur photographers (and a few curious police officers) and set off at night into the bowels of the city, literally armed with *Blitzlicht* (to produce the flash, he needed to fire a cartridge of *Blitzlicht* from a revolver). The citizens of the slums found this shooting party alarming. As Riis later wrote: "The spectacle of half a dozen strange men invading a house in the midnight hour armed with big pistols which they shot off recklessly was hardly reassuring, however sugary our speech, and it was not to be wondered at if the tenants bolted through windows and down fire-escapes wherever we went."

The damning images of the misery of tenement life that emerged from Riis's expeditions changed history. Using new halftone printing techniques, he published his photographs in a book, *How the Other Half Lives* (1890). It was a runaway bestseller. Riis traveled across the United States, narrating a show that projected his photographs from glass slides onto a wall or screen. His illuminating work caused a massive shift in public opinion. His images built support for the New York State Tenement House Act of 1901, which eliminated many of the appalling living conditions that Riis had documented. His crusade inspired a new tradition of muckraking investigative journalism that would ultimately improve the working conditions on factory floors as well.

The history of flash photography reminds us that ideas come into being through networks of collaboration. And once they're unleashed on the world, those ideas set into motion long-zoom changes that are rarely confined to single disciplines. Experiments with flash photography in one century transformed the lives of millions of city dwellers in the next century.

The Lightbulb Goes Off . . . and On

While Riis was *Blitzlicht*ing hovels to get his photographs, other big thinkers and tinkerers were looking at ways to bring artificial light to the world. The most extraordinary development between the days of kerosene lamps and today's well-lit buildings and homes was surely the creation of the electric lightbulb.

The lightbulb has come to be synonymous with genius or a "Eureka!" discovery. But there was really no single "aha" moment in the story of the lightbulb. There was, however, a famous main character: Thomas Alva Edison.

Thirty-one-year-old Edison had already invented the phonograph when he took a few months off in 1878 to tour the American West, a region that was significantly darker at night than the gaslit streets of New York and New Jersey, where Edison lived and worked. In August, two days after returning to his lab in Menlo Park, New Jersey, Edison drew three diagrams in his notebook and labeled them "Electric Light." In 1879, he filed a patent application for an "electric lamp" that displayed all the main characteristics of the lightbulb we know today.

The young wizard of Menlo Park was part of a network of people who had been inventing incandescent light for eighty years. In 1802, the British chemist Humphry Davy attached a platinum filament to an early electric battery, causing it to burn brightly for a few minutes. By the 1840s, dozens of separate inventors were working on variations of the lightbulb. Between then and 1879, scientists from America, Great Britain, Belgium, France, and Russia experimented with elements such as carbon, platinum, asbestos, and

Thomas Edison and his phonograph, circa 1878. Edison famously said, "Genius is one percent inspiration and ninety-nine percent perspiration." He lived up to his own words, sweating the details with the team in his lab.

iridium, in air or vacuums. At least half of these men had already hit upon the basic lightbulb formula that Edison ultimately arrived at: by suspending a carbon filament in a vacuum, you could prevent oxidation and keep the filament from burning up too quickly.

So why does Edison get all the credit?

One reason is self-promotion. Edison was a master of marketing and publicity. He had a very close relationship with the press. He was also a master of what we now call "vaporware": he announced nonexistent products to scare off competitors. Just a few months after Edison had started work on the electric light in 1878, he told reporters from New York that he was on the verge of launching a national lighting system. What he didn't add was that the electric lights in his lab only glowed for five minutes.

Still, Edison invited the press to his Menlo Park lab to see his revolutionary lightbulb. He brought reporters in one at a time, flicked the switch on a bulb, and let each writer enjoy the light for three or four minutes—just long enough to ensure the bulb wouldn't go out—before ushering him from the room. When asked how long his lightbulbs would last, Edison answered confidently, "Forever, almost."

It was 1882 before he made good on his claim—or at least produced a lightbulb that outperformed all others. That same year, he flipped the switch on his new Pearl Street power station and supplied electric light for an entire section of Lower Manhattan. By this time, a handful of other firms were selling their own models of incandescent electric lamps. The British inventor Joseph Swan had begun lighting homes and theaters a year earlier, in 1881. Edison was part of a network, but he became its brightest light.

An illustration from the June 21, 1882 issue of Harper's Weekly *shows Pearl Street Station, the first power plant in the United States. If Edison's lightbulb was to be used in homes and businesses, it would require a source of electricity. Edison and his team had to design and build generators that could produce enough electricity, set up a network of wires and conduits to deliver the juice, and figure out how to track usage and billing.*

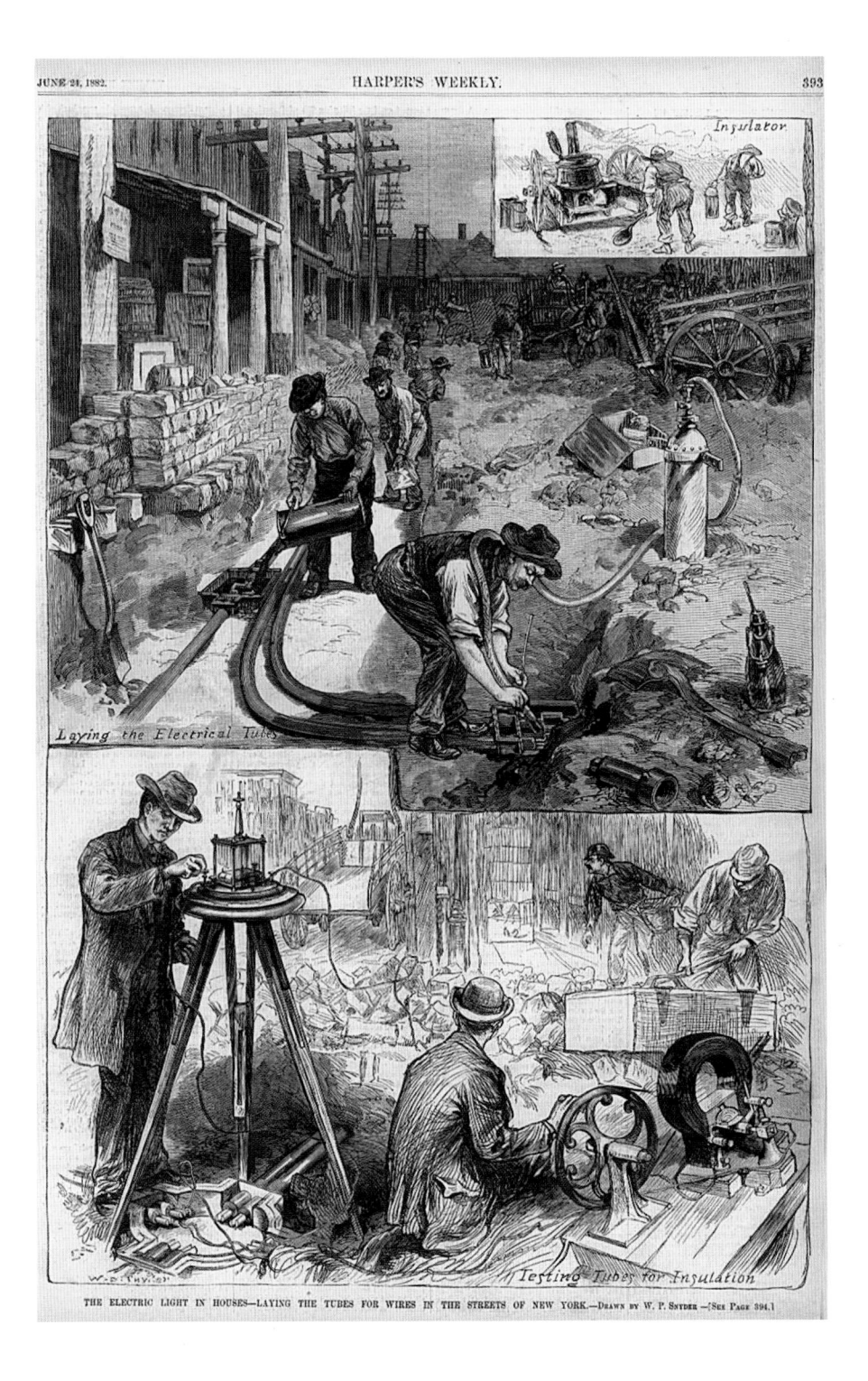

Insulator

Laying the Electrical Tubes

Testing Tubes for Insulation

W. P. SNYDER SC.

THE ELECTRIC LIGHT IN HOUSES—LAYING THE TUBES FOR WIRES IN THE STREETS OF NEW YORK.—DRAWN BY W. P. SNYDER.—[SEE PAGE 394.]

A key ingredient to Edison's success lay in the team he had assembled at his lab in Menlo Park, the "muckers." This group included a mechanic, a machinist, a mathematician/physicist, and a dozen or so other draftsmen, chemists, and metalworkers. Because the Edison lightbulb was not so much a single invention as a collection of small but ingenious improvements, the diversity of the team turned out to be an essential advantage for Edison. Menlo Park was the world's first research-and-development lab. He even paid his employees in shares, not just cash, which is a system widely used in

The first night game in Major League Baseball history, May 24, 1935. The home team Cincinnati Reds faced the Philadelphia Phillies in front of more than twenty thousand people. Before the 1930s and Edison's company, General Electric, sporting events had to be played during daylight, usually on weekdays, which meant smaller crowds.

the tech industry today. Edison didn't just invent technology; he invented an entire system for inventing, a system that would come to dominate twentieth-century industry.

Lightbulbs and electric lighting changed everything. Factories could now run twenty-four hours a day, which created more work shifts and increased productivity. Crime dropped because of electric streetlights. Lighting transformed sports and entertainment, and neon lights jazzed up signage. Private homes lit up with the flick of a switch, which eventually opened the door to other electrical appliances. The refrigerator, the washing machine, the vacuum cleaner, the food mixer—these appliances cut the hours homemakers spent on chores, freeing many women to enter the workforce.

New technology opens up new possibilities—and how we explore those possibilities is wide-open.

Las Vegas's glitzy neon-lit Strip. When neon gas is isolated and charged with an electric current, it glows. In the early 1920s, Tom Young, an enterprising sign-maker, realized he could put neon in bendable glass tubes to create some impressive signs. His idea and company took off, especially in Vegas.

Beam Those Bar Codes!

If you've ever read a superhero comic or watched a sci-fi movie, you've probably seen a villain or alien get zapped by a beam of light. The laser weapon—whether we called it a death ray or a lightsaber or a heat ray—has been a standard element in science fiction for more than a century.

Actual laser beams did not exist until the late 1950s, and they didn't become part of everyday life until the 1970s. When the laser did finally come about, its first mainstream use wasn't as a powerful weapon, but for something sci-fi authors never imagined: scanning bar codes.

A laser is a single, incredibly concentrated beam of light. It does not exist in nature; it's created by human technology. Natural or ordinary electric light has several wavelengths, or colors, which spread out. A laser beam has only one wavelength, which travels in only one direction. Like the lightbulb, the laser was not the creation of a single inventor; it was the result of work at research labs and the independent tinkering of physicist Gordon Gould, who coined the word "laser" and patented a design for one.

So lasers are very intense, narrow light beams, and they can be used with incredible precision. Hold those thoughts, and—sci-fi like—let's jump back to the 1940s and look at another invention: bar codes.

In the late 1940s, two graduate students, Bernard Silver and Norman Woodland, created a machine-readable code to identify products and prices. An early bar code looked like a bull's-eye, but required a *five-hundred*-watt bulb—almost ten times brighter than your average lightbulb—and wasn't even very accurate. The invention of the laser, though, made small handheld scanners possible. The ray gun of fiction had become a retail weapon in real life.

An illustration from an early edition of H. G. Wells's ground breaking novel The War of the Worlds *by Brazilian artist Henrique Alvim-Corrêa, 1906. Wells helped create the science fiction genre and introduced the important sci-fi weapon, the "heat ray."*

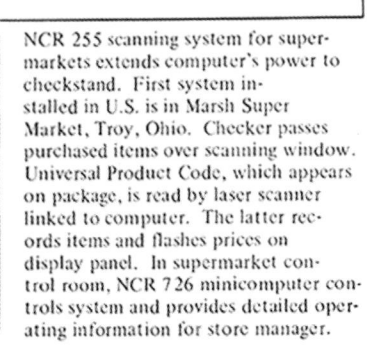

NCR 255 scanning system for supermarkets extends computer's power to checkstand. First system installed in U.S. is in Marsh Super Market, Troy, Ohio. Checker passes purchased items over scanning window. Universal Product Code, which appears on package, is read by laser scanner linked to computer. The latter records items and flashes prices on display panel. In supermarket control room, NCR 726 minicomputer controls system and provides detailed operating information for store manager.

These photos show the first product in history—a pack of chewing gum—having its bar code scanned by a laser at an Ohio supermarket, June 26, 1974.

Bar code technology spread slowly: only one percent of stores had bar code scanners as late as 1978. But scanning bar codes provided stores with information not just about price but about sales, inventory, restocking, and customer base. With more precise product and sales information, retail companies could be more savvy about how many employees and what inventory they needed. Chains and supermarkets exploded into huge big-box stores all over the world. Today, bar codes are used in many places beyond the checkout line. A QR (Quick Response) code, a type of bar code invented in 1994 in Japan, arranges black squares in a pattern on a white grid, which can be "read" by a smartphone or camera. QR codes are widely used in advertising and promotion, online shopping, and services like banking and travel. They show up on everything from billboards to business cards to T-shirts.

Laser Precision

The laser beams that first found practical use scanning bar codes have now infiltrated almost every aspect of modern life: surgery and medical procedures, manufacturing, drilling and welding, music and movies, global communications. Lasers are even going nuclear . . . hopefully.

Scientists at the National Ignition Facility (NIF) at the Lawrence Livermore National Laboratory in Northern California have built the world's largest and highest-energy laser system. Their goal is to use lasers to create a new source of energy based on nuclear fusion, the process that occurs naturally in the dense core of our sun and other stars.

At the NIF, a single pulse of low-powered laser light is sent off through fiber-optic cables into the cavernous ignition room, where it splits into 192 separate laser beams; their power is amplified a million billion times, reaching a total output of five hundred thousand gigawatts. All 192 beams fire simultaneously on a tiny bead of hydrogen. The lasers have to be positioned with mind-blowing accuracy: it's as if you were a pitcher in a San Francisco ballpark trying to strike out a batter in the Los Angeles stadium . . . 350 miles away.

When the laser beams strike it, the hydrogen releases energy. *Lots* of energy—as in, nuclear fusion, the super-hot, super-dense, super-pressured process inside stars that fuses hydrogen atoms together and releases a staggering amount of energy. For the fleeting moment that NIF lasers compress the hydrogen target, that fuel pellet is the hottest place in the solar system—hotter even than the center of the sun.

In 2013, NIF announced that the device had, for the first time, generated net positive energy during several of its shots: by a slender margin, the process created a little more energy than the lasers used up. However, not all scientists agreed on these findings, and by 2016, serious questions had been raised

as to whether NIF could ever reach its goal of fusion by laser ignition. The U.S. Department of Energy oversees NIF, and has set a 2020 deadline to determine if the agency should continue its fusion experiments or if NIF's resources and efforts should be refocused. The lasers, though, remain important research tools.

Some people might believe NIF is an expensive, glorified laser show that will never produce more energy than it takes in. Two hundred years ago, embarking on a three-year voyage into the middle of the Pacific Ocean in search of eighty-foot sperm whales might have seemed just as crazy. Yet somehow that quest fueled our appetite for light for a century. Maybe the visionaries at NIF—or another team of muckers somewhere in the world—will eventually do the same. One way or another, we are still chasing new light.

A worker inspects a massive target chamber at the National Ignition Facility, 2001.

Conclusion

····················▶

I hope the stories in this book have made you look at the world around you with fresh eyes. The next time you glance through a glass window, or flip a light switch, or check your watch to see what time it is, stop for a second to remember all of the hunches and collaborations that made those miraculous inventions so commonplace that we don't even think of them as miracles anymore. And I hope remembering that will inspire you to follow in the footsteps of the innovators you've encountered here. We live in a technologically advanced age, for sure, but that doesn't mean we've solved all the problems. There will always be new innovations to discover.

Most important innovations—in modern times, at least—arrive in clusters of simultaneous discovery. Thinking and technology come together to make a certain idea imaginable—artificial refrigeration, say, or the lightbulb—and all around the world, you suddenly see people working on the problem, and usually approaching it with the same fundamental assumptions about how that problem is ultimately going to be solved. But every now and then, some individual or group makes a leap that seems almost like time traveling. How do they do it? What allows them to see past the boundaries when their contemporaries fail to do so? The conventional explanation is the all-purpose category of "genius," of great intellectual gifts, but I suspect just as much comes out of the environment in which a genius's ideas evolve, and from the network of interests and influence that shape his or her thinking.

If there is a common thread to the inventors and discoverers who almost seem like time travelers, it is this: they worked at the margins of their official fields, or at the intersection point between very different disciplines. Working within an established field is both empowering and restricting; stay within the boundaries of your discipline and you will have an easier time making incremental improvements. (There's nothing wrong with that, of course. Progress depends on incremental improvements.) But boundaries can also become blinders, keeping you from the bigger idea that becomes visible only when you cross them. Sometimes those borders are literal, geographic ones: Frederic Tudor traveling to the Caribbean and dreaming of ice in the tropics, or Clarence Birdseye ice-fishing with the Inuit in Labrador. Sometimes the boundaries are conceptual: Ada Lovelace imagining that a calculating machine could someday be used to create music.

Steve Jobs, the great innovator of our time, talked of the creative power of stumbling into new experiences: how randomly sitting in on a calligraphy class ultimately shaped the graphic interface of the Macintosh; how being forced out of Apple at the age of thirty enabled him to launch Pixar into the business of animated movies and create the NeXT computer.

The long-zoom history of innovation shows that you don't want to be trapped by conventional wisdom, and that you need to have the commitment to stick with your slow hunches for long periods of time. You want to challenge ideas and explore uncharted territory. Make new connections rather than remain locked in the same routine. Don't be afraid of venturing into a new field just because it seems vaguely interesting, even if you find yourself disoriented at first. If you want to improve the world slightly, you need focus and determination; but if you want to make brilliant leaps forward—well, in that case, you need to get a little lost on the journey.

Acknowledgments

One of the fringe benefits of making a living by creating things—in my case, books and television shows—is that you sometimes stumble across surprising new audiences for your work once it gets into circulation. When we were making the television version of *How We Got to Now* for PBS and the BBC, we consciously thought about attracting a younger audience for our show. But our definition of "younger" was shaped by the usual viewership of history documentaries—basically anyone who wasn't yet of retirement age.

But then something surprising happened. In the weeks and months after the show began airing, I started to hear from families (and later from schools) that had enjoyed the show with their kids or grade school students. We had gone into production hoping to find a way to Generation X viewers, but it turned out we had made a show that also worked for ten-year-olds. And that is why I was so delighted when Ken Wright from Penguin reached out about the idea of adapting *How We Got to Now* for a middle grade audience. The kids who had gravitated to the TV version would have a book to call their own as well.

Writing for a younger audience was not something I had any experience with, so thankfully Ken brought on the gifted Sheila Keenan to lead the adaptation of the adult book. Catherine Frank did a masterful job running a complex editorial operation, and Jim Hoover brought a wonderful design to the book that echoes the visual flair of the original series. I'm also grateful to Ryan Sullivan and Janet Pascal, and the rest of the team for their contributions to this project, and I hope we can get the band back together for another book soon.

I should also thank some of the talented people behind the original series and adult book, starting with my television partner and producer, Jane Root, and the extended Nutopia family: Peter Lovering, Phil Craig, Diene Petterle, Julian Jones, Paul Olding, Nic Stacey, Jemila Twinch, Simon Willgoss, Rowan Greenaway, Robert MacAndrew, Miriam Reeves, Jack Chapman, Gemma Hagen, Helena Tait, Jenny Wolf, and Kirsty Urquhart. At PBS/CPB/OPB, I'm indebted to the extraordinary support from Beth Hoppe, Bill Gardner, Dave Davis, and Jennifer Lawson, along with Martin Davidson at the BBC. On the publishing side, a special thanks to my friends at Riverhead, Geoffrey Kloske, Courtney Young, and Katie Freeman, as well as to my longtime collaborator and agent, Lydia Wills.

Finally, I am beyond grateful for the support and camaraderie of my wife, Alexa Robinson, and our three boys: Clay, Rowan, and Dean. Special thanks to Dean—who happens to be an avid reader and exactly in our target audience—for giving me extensive notes on the draft manuscript that greatly improved the final version. This one's for you, kid!

The Better Ole Club Orchestra entertains a polar bear at the National Zoo in Washington, D.C., 1925.

Notes

GLASS

A small community of glassmakers from Turkey: Willach, 30.

In 1291, in an effort: Toso, 34.

After years of trial and error . . . Barovier: Verità, 17–29.

"pores or cells . . .": Braiser, 8.

"for seeing things . . .": www.technologyreview.com/s/415552/the-puzzle-of-brueghels
    -paintings-of-telescopes/

For several generations, this ingenious device: Dreyfus, 93–106.

COLD

"In a country where at some seasons": Quoted in Weightman, Kindle location 289–290.

"fortunes larger than we shall know what to do with": Quoted in Weightman, Kindle location 330.

It featured cooling rooms full of ice: Miller, 205.

A string of shipwrecks and delayed ice shipments: Wright, 12.

"better serve mankind": Quoted in Gladstone, 34.

By 1870 . . . the southern states: Shachtman, 75.

Any meat or produce . . . that had been frozen: Kurlansky, 39–40.

"Yes, the people are going to like it.": Quoted in Basile, 117.

On Memorial Day weekend: www.filmjournal.com/content/forever-74-degrees-how
-movie-theaters-keep-cool-during-summers-scorching-months

Swelling populations in Florida, Texas: Polsby, 80–88.

SOUND

Reznikoff's theory is that Neanderthal communities: www.musicandmeaning.net/issues
/showArticle.php?artID=3.2

That inventor was Alexander Graham Bell: Mercer, 31–32.

"the Engine might compose . . .": Quoted in Toole, 170.

Working out of his home lab: Hijiya, 58.

As a transmission device for the spoken word: Thompson, 92.

"much of the power . . .": wyntonmarsalis.org/news/entry/on-martin-luther-kings-legacy

Just a few days before this catastrophe: Frost, 466.

The German U-boats (submarines) roaming the North Atlantic: Frost, 476–477.

CLEAN

"first rule of nursing" and "keep the air . . .": Nightingale, 8.

When almost a hundred thousand . . . a lot of excrement: Chesbrough, 1871.

Many city authorities subscribed: Miller, 123.

Building by building, Chicago was lifted: Cain, 357.

Within three decades, more than twenty cities: Burian, Nix, Pitt, and Durrans, 33–62.

Worked in the Vienna General Hospital: Goetz, Kindle location 612–615.

Koch established a unit of measure: McGuire, 50.

Leal's interest in public health: McGuire, 112–113, 144.

In 1916, Annie Murray: The Clorox Company, 18–22.

In 2011, the Bill and Melinda Gates Foundation: www.gatesfoundation.org/What-We-Do
/Global-Development/Reinvent-the-Toilet-Challenge

TIME

To confirm his observations: Kreitzman and Foster, 33.

The watchmakers were the advance guard: Mumford, 134.

"The marvelous property of the pendulum": Drake, Kindle location 1639.

"On a rainy day": Thompson, 71–72.

Dennison's "Wm. Ellery" watch . . . cost just $3.50: Priestly, 21.

"the day of two noons": Bartky, 41–42.

In the 1890s . . . Marie Curie proposed: Senior, 244–245.

LIGHT

When darkness fell, they drifted: Ekirch, 306.

In the deep waters of the North Atlantic: Dolin, Kindle location 1272.

The candle business became so lucrative: Dolin, Kindle location 1992.

In just over a century, nearly three hundred thousand: Irwin., 51–52.

"Forever, almost.": Quoted in Sides, 178.

In October 1887, a New York newspaper . . . *Blitzlicht*: Riis, Kindle location 2228.

"The spectacle of half a dozen": Riis, Kindle location 2238.

His images built support for: Yochelson, 148.

Bibliography

Bartky, I. R. "The Adoption of Standard Time," *Technology and Culture* 30 (1989).

Basile, Salvatore. *Cool: How Air Conditioning Changed Everything*. Fordham, 2014.

Braiser, M. D. *Secret Chambers: The Inside Story of Cells and Complex Life*. Oxford, 2012.

Burian, Steven J., Stephan J. Nix, Robert E. Pitt, and S. Rocky Durrans. "Urban Wastewater Management in the United States: Past, Present, and Future," *Journal of Urban Technology* 7, no. 3, 2000.

Cain, Louis P. "Raising and Watering a City: Ellis Sylvester Chesbrough and Chicago's First Sanitation System," *Technology and Culture* 13, no. 3 (1972).

Calabro, Marian. *The Clorox Company: 100 Years, 1,000 Reasons*. The Clorox Company, 2013.

Chesbrough, E. S. "The Drainage and Sewerage of Chicago," paper read (explanatory and descriptive of maps and diagrams) at the annual meeting in Chicago, September 25, 1887.

Dolin, Eric Jay. *Leviathan: The History of Whaling in America*. Norton, 2007. Kindle.

Drake, Stillman. *Galileo at Work: His Scientific Biography*. Dover, 1995. Kindle.

Dreyfus, John. *The Invention of Spectacles and the Advent of Printing*. Oxford University Press, 1998.

Ekirch, A. Roger. *At Day's Close: A History of Nighttime*. Phoenix, 2006.

Frost, Gary L. "Inventing Schemes and Strategies: The Making and Selling of the Fessenden Oscillator," *Technology and Culture* 42, no. 3 (2001).

Gladstone, John. "John Gorrie, the Visionary. The First Century of Air Conditioning," *The Ashrae Journal*, article 1 (1998).

Goetz, Thomas. *The Remedy: Robert Koch, Arthur Conan Doyle, and the Quest to Cure Tuberculosis*. Penguin, 2014. Kindle.

Hijiya, James A. *Lee de Forest and the Fatherhood of Radio*. Lehigh University Press, 1992.

Irwin, Emily. "The Spermaceti Candle and the American Whaling Industry," *Historia* 21 (2012).

Kreitzman, Leon, and Russell Foster. *The Rhythms of Life: The Biological Clocks That Control the Daily Lives of Every Living Thing*. Profile Books, 2001.

Kurlansky, Mark. *Birdseye: The Adventures of a Curious Man*. Broadway Books, 2012.

Marsalis, Wynton. "On Martin Luther King's Legacy." January 16, 2012. wyntonmarsalis.org/news/entry/on-martin-luther-kings-legacy.

McGuire, Michael J. *The Chlorine Revolution*. American Water Works Association, 2013.

Mercer, David. *The Telephone: The Life Story of a Technology*. Greenwood, 2006.

Miller, Donald L. *City of the Century: The Epic of Chicago and the Making of America*. Simon & Schuster, 1996.

Mumford, Lewis. *Technics and Civilization*. Routledge, 1934.

Nightingale, Florence. *Notes on Nursing*. Harrison, 1879.

Polsby, Nelson W. *How Congress Evolves: Social Bases of Institutional Change*. Oxford University Press, 2005.

Priestley, Philip T. *Aaron Lufkin Dennison—an Industrial Pioneer and His Legacy*. National Association of Watch & Clock Collectors, 2010.

"The Puzzle of Brueghel's Paintings of Telescopes," *MIT Technology Review*, October 2, 2009. www.technologyreview.com/s/415552/the-puzzle-of-brueghels-paintings-of -telescopes/

Riis, Jacob A. *How the Other Half Lives: Studies Among the Tenements of New York*. Dover, 1971. Kindle.

Senior, John E. *Marie and Pierre Curie*. Sutton Publishing, 1998.

Shachtman, Tom. *Absolute Zero and the Conquest of Cold*. Houghton Mifflin, 1999.

Sides, Hampton. *In the Kingdom of Ice*. Doubleday, 2014.

Thompson, E. P. "Time, Work-Discipline and Industrial Capitalism," *Past & Present* 38 (1967).

Toole, Betty Alexandra. *Ada, the Enchantress of Numbers: Poetical Science*. Critical Connection, 2010.

Toso, Gianfranco. *Murano Glass: A History of Glass*. Arsenale, 1999.

Verità, Marco. "L'invenzione del cristallo muranese: Una verifica analitica delle fonti storiche," *Rivista della Stazione Sperimental del Vetro* 15 (1985).

Weightman, Gavin. *The Frozen Water Trade: How Ice from New England Lakes Kept the World Cool.* HarperCollins, 2003. Kindle.

Willach, Rolf. *The Long Route to the Invention of the Telescope.* American Philosophical Society, 2008.

Wright, Lawrence. *Clean and Decent: The Fascinating History of the Bathroom and the Water Closet.* Routledge & Kegan Paul, 1984.

Yochelson, Bonnie. *Rediscovering Jacob Riis: The Reformer, His Journalism, and His Photographs.* New Press, 2008.

Recommended Resources

BOOKS

Bridgman, Roger. *1,000 Inventions and Discoveries*. DK/Smithsonian, 2014.

Ignotofsky, Rachel. *Women in Science: 50 Fearless Pioneers Who Changed the World*. Ten Speed Press, 2016.

Jones, Charlotte Foltz. *Mistakes That Worked: The World's Familiar Inventions and How They Came to Be*. Delacorte Books for Young Readers, 2016.

Macaulay, David. *The Way Things Work Now*. Houghton Mifflin Harcourt Books for Young Readers, 2016.

ONLINE RESOURCES

PBS: *How We Got to Now*

www.pbs.org/how-we-got-to-now/home

Smithsonian's National Museum of American History/Lemelson Center for the Study of Invention and Innovation

invention.si.edu

TED Talks to Watch with Kids

www.ted.com/playlists/86/talks_to_watch_with_kids

Index

•••••••••••••••► Entries in *italics* indicate illustrations.

Photo Credits

Every effort has been made to trace copyright holders and to obtain their permission for the use of copyright material. The publisher apologizes for any errors or omissions and would be grateful if notified of any corrections that should be incorporated in future reprints or editions of this book.